From Print to Electronic

The Transformation of Scientific Communication

by Susan Y. Crawford,
Julie M. Hurd, and Ann C. Weller

ASIS Monograph Series

Published for the
American Society for Information Science by

Information Today, Inc.
Medford, NJ
1996

Library of Congress Cataloging-in-Publication Data

Crawford, Susan Y.
    From print to electronic : the transformation of scientific
communication // by Susan Y. Crawford, Julie M. Hurd, and
Ann C. Weller
        p.   cm.
    Includes bibliographical references and index.
    ISBN 1-57387-030-7 (hardcover)
    1. Communication in science.. 2. Telecommunication in science.
3. Electronic publishing   I. Hurd, Julie M. II. Weller, Ann C.
III. Title.
Q223.H87   1996
501'.4--dc20   96-26791
                                                                            CIP

Price: $39.50

Published by: Information Today, Inc.
                           143 Old Marlton Pike
                           Medford, NJ 08055-8750

Distributed in Europe by: Learned Information Ltd.
                                   Woodside, Hinksey Hill
                                   Oxford OX1 5AU
                                   England

The opinions expressed by the author of this book do not necessarily reflect
the position or the official policy of the American Society for Information
Science.

Book Editor: James H. Shelton
Cover Design: Jeanne Wachter

Printed in the United States of America

Contents

Foreword

Electronic Media and Technical Communication: Values, Uses, Technical Environments, and Dimensions of Media

The point of this book is that we have achieved huge and generalized storage and communication capabilities, and that one very important aspect is the use of electronic media for information transfer in science and technology. The authors are among the first voices that have no major stake in the information field who have selected areas of great importance, who have developed some substantial data, and who have furnished the first suggestions of the need for quality scholarship in what has been a literature that is, virtually, without standards.

In this Foreward, I attempt to provide some background for the development of electronic media within the context of what scholars have learned, including Merton's over sixty years of work on scientific communication.

Inventing Media for Scientific Communication

Science originally emerged as a form of very serious and highly competitive activity with much concern for discovery, but little concern with communication. Henry Oldenberg, in assuming responsibility for publication of the *Philosophical Transactions* of the Royal Society in the seventeenth century, effected what may be regarded as the major social engineering feat in the history of science. He created a scientific journal that would move news of discovery to scientists in a reasonably quick way, that was openly available, and that assured authors of enduring identification with their discoveries. The critical feature became communication, which was not present in earlier views of science, but which is now regarded as a central element in scientific work.

By the end of the eighteenth century, the military value of science was recognized, and later in the nineteenth century, its commercial importance. The overwhelming increase in scientific publications fostered a number of early attempts to cope with the flood of literature, such as compression of papers into abstracts and their compilation into abstract journals. The next great achievement would be speed and ease of access, not compression.

Crawford, Hurd, and Weller have taken on the great challenge of exploring the question of whether electronic media can replicate functions of the traditional media for scientific communication and replace, as examples, scientific meetings, technical journals, reviews and proceedings. The authors also consider the development of very large databases that exist only because of new computer and communications technology and that create new social mechanisms in science and technology.

Issues and Concerns

Great concern and sensitivity are, of course, required in developing more complex communication devices that impinge on the behaviors of scientists and engineers. For a perspective on technical communication, I would like to suggest some caution and the need to distance ourselves.

America has had an enviable history of believing that information availability is essential to a healthy and democratic society. This tradition may run counter to competitive forces in the practice of science where creative people are not always willing to spill their guts to the world as a whole.

We can look at technology for decades and not realize its ultimate effects. Who could have foreseen the effect of the telephone, the automobile, air travel, and television on such seemingly established areas as the family, the city, crime, and government? Similarly, we seem to take for granted that increases in the most explicit capabilities—speed, storage, and comprehensiveness—of information systems will improve human functioning and well being.

Other issues concern the inherent fragility of databases because of their total dependence on high technology, the rapid obsolescence of data in most hot fields, and the view that more and faster information is always better. In use of data, creativity and original discovery in science should be distinguished from secondary analysis in technology. And the flood of information on the Internet will likely drive scientists to even higher levels of elitism. Finally, there is the undervaluation of paper as a convenient and perhaps ideal medium that can be used under very difficult environmental conditions.

As a student of scientific communication and of the structures of scientific literatures, I have been grateful to review this book. The authors have undertaken a difficult task and acquitted it with distinction.

Belver C. Griffith
College of Information Science and Technology
Drexel University, Philadelphia

Preface

Many scientific information specialists over the past 30 years have benefited from the research of social scientists William Garvey and Belver Griffith. The team of Garvey and Griffith and their co-workers first mapped, in the 1960s, the steps in the scientific communication process from inception of a research project through the dissemination of results. Their work spanned a decade and established the baseline from which we have measured changes in scientific communication. When their three-volume work was published, it was an important milestone in bringing together and in quantifying the diffuse strands of the scientific communications system. The fact that Garvey and Griffith's papers on scientific communication continue to be regularly cited is a reflection of their enduring contribution to information science.

In this analysis of the shift from print to electronic communication systems, we have selected the Garvey/Griffith model as a starting point from which we view changes and project future developments. This work grew out of a conversation at an American Society for Information Science meeting in late 1993. Our shared appreciation for the insights provided by Garvey and Griffith led us to speculate on how the model they developed might be updated to reflect the changes brought about by emerging information technologies. We began to explore the transformation of scientific communication during a time when initiatives such as the National Information Infrastructure, the World Wide Web, digital libraries, and electronic journals made possible electronic alternatives to familiar scientific journals, conferences and other means of sharing information.

In this monograph we have focused on scientific specializations that have adopted and made creative use of some aspect of technology to better communicate the findings of research. We studied exemplars—space sciences, high energy physics, and human genome research—that are employing technology to devise new and better ways to collect and disseminate scientific information. We believe that these specialties may be prototypes of future developments that will spread to other disciplines and eventually transform the entire scientific communication system. Even as we worked on this book we saw changes underway that caused us to rethink and rewrite. It is very clear to us that still other develop-ments, as yet unknown, will add new dimensions to our descriptions and new facets to the models we present here as this scenario continuously unfolds.

From 1994-1996, we visited many sites and consulted with numerous scien-tists. This was necessary to obtain a perspective, as new systems were developed or changed so fast that we could not rely on published reports. We queried sci-entists at meetings and through electronic mail, keyed into home pages for the latest versions of position papers and policy statements, and visited research cen-ters for interviews. We are especially indebted to William Garvey and Belver Griffith for providing the foundation for this book. The authors also gratefully acknowledge the help of many, including the following scientists and colleagues.

Harvard-Smithsonian Center for Astrophysics, Cambridge, Massachusetts:
> Brian Marsden, Associate Director, Planetary Sciences who gave a good overview of the field yesterday and today; Guenther Eichhorn who reviewed the developing ASTROBROWSE system; Daniel Green who explained the role of the Central Bureau for Astronomical Telegrams; and Michael Kurtz who described the ADS abstracting Service. Joyce Watson was especially helpful in explaining the major information systems in astronomy and astrophysics.

Northwestern University, Evanston, Illinois:
> Lloyd Davidson, Science Librarian and Robert Michaelson, Head, Engineering Library; Giles Novak and Guy Miller, Department of Astronomy

National Aeronautics and Space Administration, Goddard Space Flight Center, Greenbelt, Maryland–Astrophysics Data Facility:
> Invaluable information was provided by staff on the NASA mission and on National Space Science Data Center operations: James L. Green, Chief, Space Science Data Operations; James Thieman, Head, Interoperable Systems Office; and Michael Van Steenberg, David Leisawitz, Robert M. Candey, and Cynthia Cheung.

American Astronomical Society: Peter Boyce

Los Alamos National Laboratory: Paul Ginsparg

National Center for Biotechnology Information: Barbara A. Rapp

National Center for Human Genome Research: David Benton

Stanford Linear Accelerator Laboratory Library: Hrvojve Galic

University of Chicago Yerkes Observatory Library: Judith Bausch

University of Illinois Community Systems Laboratory: Laura Shoman

University of Michigan Department of Space Research: Robert Clauer

Washington University, St. Louis, Genetics Department: Daniela Gerhard

William H. Welch Medical Library, Johns Hopkins University: Kerryn A.
> Brandt

About The Authors

Susan Y. Crawford, Ph.D.

Susan Crawford holds four degrees in the biological/social sciences and in information science, including a doctorate from the University of Chicago. She has authored or edited 134 books or papers, served on the editorial boards of nine journals, and was editor of the *Bulletin of the Medical Library Association* for eight years. Research activities include communication among scientists, biomedical communication, and statistical surveys of health science libraries. She taught at Columbia University (New York) and more recently was professor and director of the Library and Biomedical Communications Center at Washington University School of Medicine, St. Louis. Crawford has received over twenty national honors, among them Fellow of the American Association for the Advancement of Science, member of the Board of Regents of the National Library of Medicine, member of the Board of Overseers of Tufts University, and distinguished member of the Academy of Health Information Professionals. She is recipient of both the Distinguished Award and Distinguished Graduate Medal from the University of Toronto. The Medical Library Association has conferred upon her the Eliot and the President's Awards, Special Award to the Editor, life fellowship, and the Noyes Award, the Association's highest honor.

Julie M. Hurd

Julie Hurd has been an active member of the American Society for Information Science for many years on the local and national levels. She currently serves as ASIS Deputy Chapter Assembly Director and participates in program planning for SIG/STI (Scientific and Technical Information). She is also active in the Science and Technology Section of the American Library Association's Association of College and Research Libraries. She holds a Ph.D. in theoretical chemistry from the University of Chicago and an M.A. in library science from the same institution. She has worked in science libraries at the University of Chicago, Michigan State University, and the University of Illinois at Chicago, and has served as a library school faculty member at the University of Chicago and at Rosary College. She is currently head of the Science Library at the University of Illinois at Chicago. Her research interests in scientific communication and scientists' use of information grew out of her academic training in chemistry and are reflected in various journal articles, book chapters, and conference presentations.

Ann C. Weller

Ann Weller is Deputy Director at the Library of the Health Sciences, University of Illinois at Chicago. She was previously head of the Reference Department at the Library of the American Medical Association. Her extensive committee appointments within the Medical Library Association (MLA) have focused on research and scholarship. She was a member of the Research Task Force that developed the MLA policy statement *Using Scientific Evidence to Improve Medical Practice*, and is currently serving on the task force charged with its implementation. She also chaired the association's Committee on Grants and Scholarships and is presently on the Editorial Board of the *Bulletin of the Medical Library Association*. Her research on editorial peer review was reported at the first two International Congresses on Peer Review in Biomedical Publications and published in both the library and medical literature. Weller received her M.A. in Library Science from the University of Chicago and is currently a doctoral candidate at that university.

Chapter One

Scientific Communication and the Growth of Big Science

Susan Y. Crawford

That science and technology are central to economic competitiveness and national survival is a core concept of American policy today (Technology for America's Economic Growth 1993). However, in the volatile world of economic and political priorities, this was not always so. The notion that science could transform America gained impetus in the late nineteenth century with exciting, new discoveries in the physical and chemical sciences (Kargon, Leslie, and Schoenberger 1995). Then, corporations such as AT&T and General Electric set up specialized laboratories that pursued goal and product-directed research. From these beginnings, the modern infrastructure for scientific research emerged—a collaboration of industry, academia, and government. World War II accelerated the trend through major contributions of science to national defense—conversion to wartime industry and development of powerful weaponry culminating in the creation of the atomic bomb. In 1945, the foundation for federal support in succeeding years was laid by Vannevar Bush (1945), a leader of wartime science, in his report "Science—the Endless Frontier." Bush persuasively argued that basic research was essential to advancement in three important areas of American life—defense, industry, and health; that the federal government should assume responsibility for its support; and that research should be under the control of independent scientists.

Federal funding for science continually increased after the war years, especially during the first decade and a half, a time of prosperity, economic growth, and optimism. In 1950, the National Science Foundation was established. At the beginning of the Cold War, competition with the Soviet Union on many

1

fronts spurred greater scientific efforts, exemplified by space exploration and projects such as landing a man on the moon. Total spending on research and development increased by 42 percent in inflation-adjusted dollars between 1960 and 1968 (Dickson 1988). During the same period, government support for basic research increased three-fold and universities, which had rejected almost all public funding before the second World War, were the major recipients (ibid).

By the middle of the 1960s, the Vietnam War and subsequent social unrest plagued the Johnson Administration and priorities began to shift toward social programs and civil rights. The payoff from basic research was questioned and applications research was now emphasized. During Nixon's term, Research Applied to National Needs (RANN) was introduced and both political support and funding for the basic sciences were cut. The economic recession that followed in 1973-74 brought the realization that, in a global economy, continuous technological innovation was necessary for the United States to compete successfully. Basic research was again viewed as important to the American economy, characterized by President Carter as an "investment" rather than an overhead expenditure (ibid). The federal research budget subsequently took a great leap. During the Reagan administration, support for military and for industrial research had high priority, while spending for the social sciences was cut by some 50 percent. With the rebound of industrial output in Japan and in Western Europe, American businesses focused on reducing the costs of production rather than on long-term research and development that delivered uncertain payoffs. To industry, scientific research had become a bottom line strategy. Even now, early in 1996, profound changes in the organization of research and development are taking place, a result of the new Republican-majority Congress elected in November 1994 (Mervis 1995).

As Hurd (1996) reported, Big Science has become "Bigger Science," as shown in data from the National Science Foundation. Scientists in the United States currently number some 700,000, an increase by almost a factor of six since World War II. The Hubble telescope cost about $2 billion to develop and the Superconducting Supercollider, that was discontinued in 1994, had planned to include 500 scientists holding Ph.D.s on a single experiment (Galison 1992). But as Hevly (1992) points out, the creation of Big Science does not reflect simply changes in scale. New forms of institutional, political, and social organizations arose, among them, the concentration of resources in fewer centers and in those dedicated to specific goals; specialization of workers and a hierarchical structure; and the attachment of political and social significance to scientific projects (Hevly 1992). Silicon Valley is viewed by Kargon and co-authors (1995) as the archetype of the modern university-industry-government nexus for scientific research and development.

2

Policies of the Clinton-Gore Administration underscore the value of R&D for putting the economy on track, for increasing employment and for improving the quality of life. As a corollary, scientific information and its transfer are also assigned important roles. A communications infrastructure that would provide instant access to information was deemed necessary and feasible and the "national information superhighway" was conceived (Information Infrastructure Task Force 1993; President Clinton 1994).

The scientific communications system is a product of research and its application and, as such, shares the same environmental determinants. The system had evolved gradually from the seventeenth century, with the emergence of scientific disciplines and their journals published by newly-formed professional societies (Kronick 1962). In the post-World War II era of great social change, Big Science and powerful new technology have clearly modified the way information is produced, managed, and used. Whether public or private-sector based, research had become a collaborative venture with long-term goals, large budgets, costly laboratories, and socioeconomic consequences. The communications systems in each of the disciplines, specialties, or problem areas have evolved through new developments in the media, integration of resources, expansion of telecommunications networks, and application of computers to manage information and to extend cognition.

The purpose of this monograph is to provide an overview and to examine these changes against existing models of scientific communications. Succeeding chapters provide case studies of the impact on the communications systems of three specialty areas: the Human Genome Project, the space sciences, and high energy physics.

The Garvey/Griffith Model of Scientific Communication

Price (1968) has calculated that, since the 1700s, as measured by the number of journals founded, the output of scientific information has grown exponentially. The devastating impact of the current "bigness" of science on the individual scientist's ability to cope was noted. As scientists struggled to manage this information and wondered how to sift through the burgeoning volume to find what they wanted, the computer appeared.

In the early 1960s, William Garvey at Johns Hopkins University and Belver Griffith at the American Psychological Association suggested that, to understand and to optimize the communication process, it is necessary to examine it within the context of specific disciplines. They did just that, focusing on the field of psychology, and mapped steps in the scientific communications system from the inception of a project through the dissemination of results (Reports of the American Psychological Association's Project 1963, 1965, 1969).

3

Garvey and Griffith found that, on the average, almost four years elapse from the time research begins through journal publication, and another year before the results are entered into abstracting and indexing services. And nearly a decade passes before the findings are incorporated into textbooks and encyclopedias as established knowledge!

Communications Model of the 1990s: the Digital Revolution

The Garvey/Griffith model was developed before the use of information handling technologies that are commonplace today. Some thirty years later, emerging technologies have altered and enhanced options for managing and communicating information that were undreamed of earlier. How have these changes affected the Garvey/Griffith model that provided the standard for so many years?

Space stations, scientific laboratories, and biological repositories generate and collect huge quantities of data, text, and images in digital format. Their transfer over national and international communications networks has enabled the building of large, specialized machine-readable databases. For many specialties, the data are collected, reviewed, and electronically deposited for use prior to publication. In the Human Genome Project, mapping data are received by Johns Hopkins University from laboratories around the world and screened by a committee before becoming part of the Genome Data Base (GDB). When a scientist enters data into the GDB, a sequential number is assigned to establish priority and, at the same time to indicate that the data have been validated. Increasingly, biological journals will not publish a paper unless a sequential number has been assigned by GDB (Crawford, Stucki, and Halbrook 1993). Publication in a scientific journal usually follows the deposit of information. Sequence information is coordinated by the National Center for Biotechnology Information—U.S. National Library of Medicine from global databases in the United States, Europe, and Japan.

Retrieval is facilitated by electronic links between data, between information (relationships among data), and between knowledge (meaning of the data), as demonstrated in the genome database (Matheson 1995). Access is through the omnipresent personal computer, which is now routinely used in scientific research.

Electronic Invisible Colleges and Collaboratories

Throughout the world, at the frontiers of most problem areas in science, there are usually individuals or groups of scientists pursuing common objectives or working on related problems. Those arriving at the research front find

4

others, with similar basic training in specialty and methods, looking at the same problems. Although the players are rather a mobile lot, moving from one problem to another during their lifetimes, they tend to know each other, to talk to each other, to be aware of each other's work, and to cite each other (Price 1971).

Relationships are established in part through the federal granting system that puts the project in the public domain at the time of submission. Out of this has emerged social structures that Price has called "invisible colleges," communities defined by communication ties and by collaboration stemming from related or shared areas of investigation. Early studies of the social organization and the diffusion of ideas among such research communities were made by Crane (1972) in rural sociology, Crawford (1971) in sleep research, and Mullins (1968) in biology under the mentorship of Price.

In the digital era, informal communication is much different from the Garvey/Griffith model. Telephone conversations, visits to laboratories, and encounters at meetings are supplemented by computer-to-computer communications on national and international networks. Scientists regularly converse and transmit data electronically. Through electronic conferences, these researchers seek advice, consensus, and new methods. A message or questionnaire can be quickly relayed to a wide audience and subscribers to listserves receive newsletters, discussions, and critiques from their colleagues. Meetings are televised and key addresses or papers are available through the Internet. The invisible college has become electronic.

An organization or consortium can now encode the knowledge base of communities of scientists or invisible colleges and build a software environment to manipulate this knowledge from laboratories. Schatz (1992) and co-workers have built what they call an "electronic community system" for scientists studying the nematode worm *C . Elegans*. The system includes archival literature, "fundamental" items of data for the community (as found in their database management systems), and intermediate news (as found on electronic bulletin boards). Such systems are also being developed in other research communities (e.g., Human Genome Project, brain research, structural biology), in institutions (e.g., hospitals and academic health science centers), and in industry (Hamilton 1992; Gibbons 1992). In 1994, the National Science Foundation awarded funding to six organizations to develop "collaboratories," in which scientists can interact electronically to share data, instrumentation, and computational resources (National Research Council 1993).

The basic model for information access and distribution among digital systems, Lynch has noted, involves obtaining a copy of a file on one's personal machine, altering or adding to it as one wishes, and making it available to others. A key advantage is the capability for manipulating files from word processors and desktop publishing systems (Lynch 1994). While this emerging "publication on demand" system has advantages, many difficulties remain to be resolved, as will be shown in our examination of several systems in the succeeding chapters.

Discussion

Responding in large part to economic and political forces, the scientific enterprise has burgeoned during the past fifty years from "Big Science" to "Bigger Science." The scientific communications system, which provides the framework around which science advances, shares the same environmental determinants as Big Science. When research results exceeded the ability to retrieve them—a low point in the 1960s—scientists half seriously joked that one could easier redo an experiment than find reports of the work among scientific papers. This sentiment reflected the dire conditions at the time that Garvey and Griffith provided their important natural history of the origination, transfer, and use of scientific information. It was also a plea for federal support of information retrieval.

The conditions that gave birth to the journal and the book, and that made each in its time the ideal medium for communicating, have changed. In the digital world, information conveyance devices that provide identification, transmission, and storage functions are no longer discrete physical entities. The "virtual library," that focuses on access, transcends publications, walls, and institutions. What Garvey and Griffith eloquently stated in the 1960s, Humphreys and Lindberg reiterate today: the crucial element is users' needs—relevance of the information, economy of effort in getting it, and facilitating conceptual connections (Humphreys and Lindberg 1994).

Today, we have the technological potential for designing systems that meet those needs. In the succeeding chapters, we examine the methods and the problems of innovative systems developed in three research areas. We begin with the original Garvey/Griffith model and, based upon technological and socioeconomic assumptions, we develop several possible outcomes. Weller studies information management in the international effort to map the human genome. The space sciences, which continuously collect data in real time and online from both airborne and ground-based facilities, pose formidable problems, as reported by Crawford. Hurd examines the fast-moving field of high energy physics, which epitomizes Big Science. From the data, the authors examine changes in the scientific communications system and make projections that reflect the digital age, which promises ever increasing and significant additions to our storehouse of scientific knowledge.

References

Bush, V. 1945. *Science: the Endless Frontier.* Washington: Government Printing Office.

Crane, D. 1972. *Invisible Colleges; Diffusion of Knowledge in Scientific Communities.* Chicago: University of Chicago Press.

Crawford, S. 1971. Informal communication among scientists in sleep research. *J. American Society for Information Science*, 22:301-310.

Crawford, S., L. Stucki, and B. Halbrook. 1993. Managing information in biomedical research: the Human Genome Project. In N.C. Broering, (ed.) *High Performance Medical Libraries.* Westport, CT, Meckler.

Dickson, D. 1988. *The New Politics of Science.* Chicago: University of Chicago Press.

Galison, P. 1992. The many faces of science. In P. Galison and B. Hevly, eds. *Big Science: The Growth of Large Scale Research.* Palo Alto: Stanford University Press.

Gibbons, A. 1992. Databasing the brain. *Science* 258:1872-73.

Hamilton, D.P. 1992. Coping with an "embarrassment of riches." *Science* 255:397

Hevly, B. 1992. Reflections on big science and big history. In P. Galison and B. Hevly, eds. *Big science and the growth of large scale research.* Palo Alto: Stanford University Press.

Humphreys, B.L. and D.A.B. Lindberg. 1994. The UMLS project: making the conceptual connection between users and the information they need. *Bulletin of the Medical Library Association* 81(2), April:170-177.

Information Infrastructure Task Force. 1993. *National Information Infrastructure: Agenda for Action.* Washington: Government Printing Office.

Kargon, R., S. W. Leslie, and E. Schoenberger. 1995. Far beyond big science: science regions and the organization of research and development. In: P. Galison and B. Hevly, eds. *Big science: The Growth of Large Scale Research.* Palo Alto: Stanford University Press.

Kronick, D.A. 1962. *A History of Scientific and Technical Periodicals: The Origins and Development of the Scientific and Technological Press 1665-1790.* New York: Scarecrow Press.

Lynch, C.A. 1994. The integrity of digital information: mechanics and definitional issues. *Journal of the American Society for Information Science* 45(10):737-44.

Matheson, N.W. 1995. The idea of the library in the twenty-first century. *Bulletin of the Medical Library Association* 83(31): 1-7.

Mervis J. 1995 Clinton holds the line on R&D. *Science* 267:780-82.

Mullins, N.C. 1968. The distribution of social and cultural properties in informal communication networks among biological scientists. *American Sociological Review* 33:786-97.

National Research Council. 1993. Committee on a National Collaboratory Establishing the User-Developer Partnership. *National Collaboratories: Applying Information Technology for Scientific Research.* Washington, DC: National Academy Press.

President Clinton. 1994. Address before a Joint Session of Congress on the State of the Union. January 24, 1994. Washington: Government Printing Office.

Price, D. 1968. *Big Science, Little Science . . . and Beyond.* New York: Columbia University Press.

―――. 1971. Invisible college research: state of the art. In S. Crawford, ed. *Informal Communication among Scientists: Proceedings of a Conference on Current Research.* Chicago: American Medical Association.

Reports of the American Psychological Association's Project on Scientific Information Exchange in Psychology. 1963, v. 1; 1965, v.2; 1969, v.3. Washington, DC: American Psychological Association.

Schatz B. 1992. Building an electronic community system. *J. Management Information Systems* 8:87-107.

Technology for America's Economic Growth, a Direction to Build Economic Strength. 1993. Washington: Government Printing Office.

Chapter Two

Models of Scientific Communications Systems

Julie M. Hurd

Developments in information technology will bring about changes in the way in which scientists communicate informally, but it is hard to see how these changes will radically affect the kinds of interactions which are recorded.

—Blaise Cronin (1982)

Communication in science is supported by a complex, interrelated system. That system has evolved gradually since the nineteenth century when many discipline-specific scientific journals began publication as organs of newly formed scientific societies. The process of producing, organizing, and disseminating scientific information consists of both formal and informal communication and involves interactions among many different organizations. Participants comprise not-for-profit and profit-sector organizations such as universities, academic departments, libraries, professional associations, research institutes, scholarly and trade publishers, database producers, and information industry vendors, as well as more loosely structured groups known as "invisible colleges." The study of scientific communication has engaged the interests of historians and sociologists of science as well as science librarians and information scientists; the literature resulting from their research is extensive.

The Garvey/Griffith Model of Scientific Communication

William Garvey, Belver Griffith, and co-workers more than thirty years ago developed a model of the scientific communication system based on their observations of psychologists (Garvey and Griffith 1972; Garvey 1979, and references cited therein). Garvey asserted that "communication is the essence of science" and that scientific communication as a social process would lend itself to the methodology of social psychology. The Garvey/Griffith model was subsequently demonstrated to be generally applicable across both the physical and

9

social sciences. It outlines the process by which research is communicated and provides details of the various stages that encompass a time frame extending from initial concept to integration of the research as an accepted component of scientific knowledge. Although the time scale varies from one discipline to another, the essential elements of the model appear to be universal.

The Garvey/Griffith model, as postulated in numerous publications of the 1970s, was based on the communication channels then operational. These were both informal and formal and included personal (oral) communications among individuals and groups as well as publication in journals and books. Figure 2-1 provides a general representation of Garvey and Griffith's model adapted from illustrations in their publications. It outlines the communication of research findings in a typical scientific discipline.

Since Garvey and Griffith developed this model, emerging information technologies have dramatically altered and enhanced options for communicating. The application of computers to publishing has brought online bibliographic databases and large amounts of machine-readable text created to support the publication of books and journals, as well as totally electronic journals. Visionaries such as J.C.R. Licklider (1965) and F.W. Lancaster (1978) foresaw in these developments a "paperless" future. Although that future has yet to become reality, the technological foundations are in place; the economic, social, and political barriers remain to be overcome.

This chapter examines the role of emerging information technologies and explores how these may catalyze changes in the communication system. A series of models are presented that incorporate recent developments, representing new applications of information technology. We speculate, as well, on the future directions these applications suggest.

The new scenarios we describe are presented with the knowledge that past efforts to predict the future of information science have not always been accurate. Many will recall, for example, the enthusiasm among information specialists for the use of microforms that was not widely shared by scholars and scientists. Vannevar Bush's "Memex" and Fremont Rider's research library of the future both recognized potential in microformats and they offered descriptions of information systems relying heavily on utilization of this compact storage medium (Bush 1945; Rider 1944). Despite their very compelling arguments and the widespread interest generated by their writings in the scientific and information communities, microforms gained slow acceptance by information users. That resistance endures to this time and serves as a reminder of the real possibility for clouded visions of even the immediate future.

Memex has proved to be a more enduring concept and is considered by many as an early description of both a "scholar's workstation" and a "digital library." Bush described the Memex as "a sort of mechanized private file and

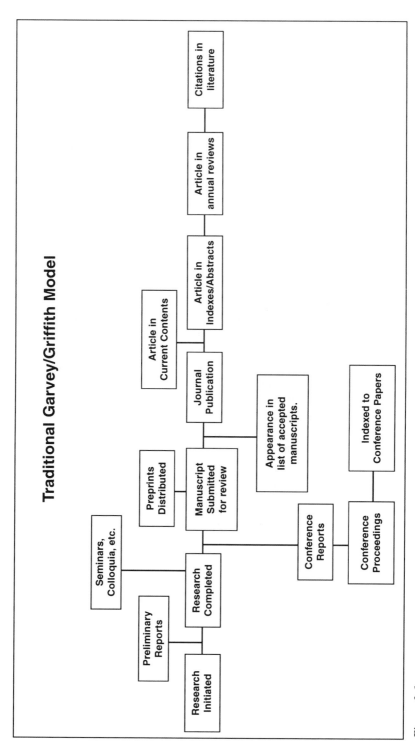

Figure 2-1

library . . . in which an individual stores all his books, records, and communications." Memex employed microfilm as the storage medium in a desk with an attached keyboard and viewing screen. Indexing of stored materials made retrieval efficient and relationships among documents could be created in a manner that sounds very much like current descriptions of hypertext links.

The technology available in 1945 did not permit construction of an actual working model of a Memex, but the potential for microfilm as a high density storage medium appealed to many readers. Bush's vision played a significant role in the efforts to utilize microformats for information storage that followed. Neither Bush nor Rider anticipated the lack of enthusiasm among scholars for the new storage medium, a reluctance to the use of microformats that has continued to this day.

Nonetheless, as did Bush and Rider, we will attempt a look into the future to new applications of technology. The models described in this chapter are offered with some confidence that they are reasonable representations of new communication paradigms. Perhaps some of them will prove to be more accurate predictions than those of early advocates of microfilm.

As we examine scientific communication and attempt to predict how it will occur in the future, one should keep in mind that not all participants in the communication process, whether individuals or organizations, share the same personal and organizational goals, sources of financial support, or reward structures. These disparities complicate setting directions and resolving problems and assure varying perspectives on controversial issues. Discussion of each model addresses these differences, poses questions that require answers, and identifies problems that must be resolved.

Computer-supported Communication as a Change Agent

The now ubiquitous personal computer, teamed with the modem and low-cost telecommunications networks, first provided new communication channels to permit computer-to-computer information transfer. Wireless technologies including satellite transmitters and cellular telephones serving as receivers now further broaden the communication base and allow emerging nations access without necessarily wiring a nationwide communication network. Classrooms, laboratories, libraries, hospitals, offices, and homes around the globe can be connected for purposes of receiving and transmitting words, data, images, and sounds. It is not surprising that, in the fall of 1995, politicians as far apart in philosophy as Vice President Albert Gore and Speaker of the House Newt Gingrich agreed on the importance of information technologies to bring about sweeping social changes.

12

Overshadowing most other technological development in its impact, both realized and potential, is the global network of networks known as the Internet. This system of interconnected computer networks has its origins in 1969 in the Advanced Projects Research Administration Network (ARPANET) that was created and funded by the U.S. Department of Defense to link four university-based supercomputers with key research sites at other academic institutions and government facilities. ARPANET functioned as a high-speed backbone to which other university and government computer networks could link to route messages with great efficiency.

The communications protocols and technical standards that continue in use today were developed by a loosely organized and dispersed community of scientist-users motivated by shared interests in scientific research. The linked networks have evolved into the Internet, now referred to in the popular media as the "Information Superhighway" and "cyberspace," and formerly as the National Information Infrastructure (NII) and the National Research and Education Network (NREN). The administration of the Internet continues to be loose; and its economic base becomes increasingly complex as commercial organizations tie in while support through government funding dwindles.

The Internet has met with general enthusiasm by almost every scholarly and scientific discipline. What's more, it has encouraged the growth among for-profit organizations of an array of services that they market to virtually all interested computer users. Although originally created to facilitate sharing of research data among a very elite community of scientists, the Internet has now broadened its user base to include every segment of society, including children in grade schools and average citizens who may use a computer for strictly recreational pursuits.

At this writing, estimates of the number of people connected to the Internet range from 16 million to 40 million, with the exact number probably unknowable. Charles Arthur, writing in *New Scientist*, suggests that arguing about the precise number of Internet users is the contemporary equivalent of earlier religious debates on how many angels could dance on the head of a pin! (Arthur 1995) It is likely sufficient to realize that the number of Internet users continues to increase and that this worldwide communications tool, with its incredible capabilities, represents a fundamental transformation in human communication.

Computer-based communication was not foreseen in Garvey and Griffith's model, but any observer of contemporary research communities could not fail to see how scientists have assimilated information technologies into their daily routines. All studies of Internet activities reveal that electronic mail is the most-used application, whether by scientists or the whole user population. The network also supports data transfer among remote sites, even extraterrestrial locations such as space telescopes and orbiting space vehicles. Opportunities to reach large audiences exist in bulletin board services and news groups as well as through listserves and computer conferences. Examples of all these and other uses are described more fully in this work to demonstrate changes in the system of communication among scientists.

In addition, the broadening of the Internet user population well beyond the original research community allows potential access to communication channels that were previously closed to many groups. The resulting democratization and popularization has the potential to bring new individuals and organizations into the system. A communication system that was originally closed to outsiders may now, because of the open nature of networks, allow glimpses into the research process and opportunities for active participation by a broader array of individuals.

The impact of universal access can only be speculated; it is premature to say whether it will actually occur. It must also be recognized that the enabling technologies are not yet available to all, even in highly developed countries, and that a significant underclass of "information have-nots" are disenfranchised because they lack adequate technology, appropriate training, or sufficient interest in the changes around them.

Recognizing the potential problem, the National Science Foundation (NSF) created a Digital Library Initiative to fund prototype digital libraries in a number of specialized disciplines with a stated goal of broadening access to knowledge. Several of the projects funded have incorporated interaction with secondary and primary educators and students and will use students to test and refine the systems that are developed. A special issue of the *Communications of the ACM* (1995) describes the digital library initiatives and explores the challenges faced by these pioneering projects.

We first examine the components of the Garvey/Griffith model and postulate how computer-based information technologies can both enhance and alter the communications system. Clifford Lynch (1993) offers a perspective employed here that distinguishes between *modernization* and *transformation* of scientific communication. Modernization is defined as the use of new technology to continue doing the same thing, but presumably in a more cost-effective and/or efficient way. Transformation is the use of a new technology to change processes in a fundamental way. At this time, we find more examples of modernization of the communication system than transformation, but that likely is attributable to the early stages of the transition. The models proposed in this chapter represent both types of responses to technological innovation.

Informal Communication

At the earliest stages of a research project, communications of findings are informal and preliminary as befitting their tentative nature. Personal communi-

14

cations within an organization are likely to occur first, with research being discussed casually in serendipitous gatherings and in more structured, but still relatively informal, research group and departmental seminars. As research progresses the investigators may also consult with colleagues at other institutions.

The existence of groups of scientists working on closely related research and communicating with each other across institutional boundaries was noted by Derek Price (1963) and Diana Crane (1972), who described the phenomenon with the term "invisible college." These informal contacts may provide information on experimental techniques or data analysis, or may clarify or validate interpretations of findings. At this stage in research, communication elicits feedback that may lead to modifications in experimental design or data interpretation.

When Garvey, Crane and others described scientific communication, the communicants relied on telephone conversations, personal correspondence, visits to other laboratories, and encounters at professional meetings, all of which could be limited by time constraints or funds available for travel. The development of electronic mail and other computer-based forms of communication, however, presents significant opportunities for informal communication, whether to known colleagues or to fellow subscribers to a particular "listserve" or discussion group.

Reports of use of these new tools and their impact on the communication structure are just beginning to appear. An article in the *Chronicle of Higher Education*, for example, offers a number of examples of scholars' use of discussion groups to facilitate informal communication about teaching and research as well as to share news on upcoming meetings, professional issues, and job opportunities (DeLoughry 1994). Perhaps such computer-based groups might be called "electronic invisible colleges;" they could be seen as representing a modernized assemblage with many of the same features described by Crane.

Blaise Cronin (1982) reviewed the role of invisible colleges in information transfer and speculated about the impact of newly-emerging communication technologies. His conclusions accurately reflect current developments:

> The invisible college is a simple yet complex telegraph system serving the needs of the scientific community. There seems to be little doubt that developments in communication technology will herald a new mode of invisible college. . . . What remains uncertain, however, is the extent to which technological innovation will encourage a broadening of the participative base and reduce the unequal distribution of benefits.

15

Numerous participants have noted that discussion groups benefit from their openness and the general lack of barriers between the well-known professors and the student members. Listserve users speak enthusiastically of the sense of collegiality that prevails. Still others cite the information they obtained promptly from knowledgeable fellow list subscribers. Whether as moderator or member, the list participant has intellectual connections to a larger community sharing the same interests; this is valued particularly by faculty at smaller or more isolated institutions for whom staying in touch with peers can be difficult and costly.

Despite these positive aspects of "being connected," not all scholars see the benefits and claim that only "lesser-known" researchers participate in list discussions. Such views may have given rise to closed lists to which subscriptions are tightly controlled in an effort to elevate the quality of discussions and screen out those not considered to be peers.

It remains unclear whether in the future of "electronic invisible colleges" open or closed groups will predominate: their role in supporting informal communication is still evolving. If "open admissions" becomes the norm, it will represent a true transformation in the communication system through the involvement of new participants formerly not part of the process.

The Role of Meetings and Conferences

As a scientific research project progresses and conclusions can be articulated with some assurance, its findings are disseminated in presentations at meetings and conferences. These gatherings vary in size and scope from small invitational meetings on well-defined topics, such as the Gordon conferences that bring together leading researchers to discuss cutting edge issues, to very large open registration professional association conferences attended by thousands. Invitations to present one's research at a conference may be made on the basis of an abstract submitted to a program committee, and thus a meeting provides an opportunity to offer very recent findings without the built-in delays of the publication process. The conference presentation may be published in its entirety in a "Proceedings," but practice varies greatly; some presentations may appear only in brief meeting abstracts, still others only as a title in a program.

Program presentations, however, are not the only reason scientists attend meetings; conferences serve multiple purposes. First, as observed above, they present the opportunity to describe one's research to an interested audience who might provide additional feedback useful in refining accounts of the research. Second, there are numerous occasions for conversations with one's "invisible college" colleagues, editors and publishers of scientific journals, planners of future conferences, and others. The formal structure of the meeting is certainly important to the attendees, but so are all the social events and opportunities for informal conversations (Oseman 1989). If video and computer conferences are to replace traditional scientific meetings, these less tangible benefits of conferences must be recognized and accommodated so that important aspects of scientific communication are not lost.

A recent experiment in computer conferencing, CHEMCONF, that incorporated many functions of the traditional conference was described at the TriSociety Symposium on Chemical Information by Thomas O'Haver, the conference organizer (O'Haver 1994). CHEMCONF, an electronic conference for chemistry educators, offered opportunities for participants to learn of new developments in their specialization, interact with others in the field, share information, and argue issues, all without traveling physically to a common location. CHEMCONF used the Internet's capabilities to transmit text and graphics of invited and contributed papers. These were distributed to all registrants according to a schedule; discussion of each paper followed through use of listserve software.

Detailed evaluations solicited from the 450 chemistry educators representing 33 countries validated the experience. Many positive comments were made concerning the content of the papers presented, but more significant were the enthusiastic endorsements of the innovative format. Participants cited the high quality of discussions following the papers: discourse that may have been facilitated by a freedom to reflect and frame commentaries in an electronic communication. Several persons with hearing and mobility disabilities were particularly appreciative of the online format. For them the traditional conference often presents barriers that diminish the quality of the experience; in an online environment they had equal opportunities for full participation. O'Haver considered his experiment a success and identified several other online conferences in chemistry that followed as further evidence supporting the approach.

More recently, the Electronic Conference on Trends in Organic Chemistry (ECTOC-1) was held on the World Wide Web (Krieger 1995). No fees were charged and access to the "papers" and "posters" was available to anyone with a Web browser such as Mosaic or Netscape. The capacity to create hypertext links supported direct contact with authors through electronic mail, and the graphical nature of the interfaces allowed inclusion of molecular and crystal structures and other images, such as photos of authors. Keyword searches of papers were supported, as well as links to older literature through a connection to the Chemical Abstracts Service database. The papers presented will be treated as published scholarly works, and those reporting new chemical information will be abstracted and added to *Chemical Abstracts*. In addition, a CD-ROM disk of the conference, including some commentary and discussion, is planned to provide an archival function. The actual papers presented will remain online for two years so that discussions can continue.

The electronic environment in which ECTOC-1 occurred supported automatic collection of very detailed statistics, which will be useful in planning future such "meetings" and analyzing the participation in this one. Its sponsors have judged it a success and are organizing for ECTOC-2.

Electronic conferences will probably never completely replace traditional face-to-face meetings. They seem destined, however, to be employed more and more to provide a complementary format that allows participants to employ scarce time and financial resources more effectively.

Formal Communication

Publication in a refereed scientific journal marks the completion of a project. When the project's findings have been reviewed and accepted by professional peers, it signifies acceptance by the scientific community and also certifies a claim on any reported discoveries. The journal serves both a communication function and an archival one in providing the record of authorship on its pages; that is why "received dates" are common on articles. It should be noted here that those working in the same specialty probably already know of research that has just been published through all the various informal communication channels previously described. The journal article, however, informs those in other specializations.

Nobel Prize-winning geneticist Joshua Lederberg, writing in *The Scientist,* affirms the continuing importance of the scientific journal in his research and reflects on the importance of the editorial process, including peer review (Lederberg 1993). He speculates on the role of informal computer-based communications and suggests changes in the publication system that he believes would enhance communication. The sort of model implicit in his discussion may be a modernized model similar to one described later in this chapter.

Suggestions for alternatives to the scientific journal are not new. The literature in library and information science as well as in scientific disciplines contains proposals dating back at least twenty years. Prior to the widespread adoption of computers in publication, these proposals tended to be based on alternative distribution schemes often involving separate articles as the basic unit, sometimes combined with use of microfilm or fiche as a distribution medium. A recent review describes some of these and assesses their minimal impact on scientific communication (Piternick 1989). Even though none of these alternatives were adopted by the scientific community, some of their features can be incorporated into newer, computer-based systems.

Despite prior dissemination of research findings through informal communication, a sense of urgency pervades formal communication, and publication delays are not acceptable to authors. A study of physicists' use of journals by the Institute of Physics identified speed of publication as one of the most important criteria an author employs in selecting a publisher (King and Roderer 1982). Although that survey was done some years ago, it is evident that authors' desire for more rapid publication has only increased. To that end, the various forms of electronic production methods available today offer much promise in reducing publication delays and facilitating scientific communication.

To be successful, electronic journals will need to address all the functions currently served by the more traditional format, including the role of publication in the reward structure for many scientists, particularly those associated with universities. A *Chronicle of Higher Education* article identifies determinants of success for the electronic journal and quotes a scholarly scientific publisher, David Rodgers of the American Mathematical Society, who observes that electronic journals "will have to offer compellingly superior capabilities to paper . . . along with lower production and distribution costs." (Wilson 1991)

More recently, the Association of American Universities, in collaboration with the Association of Research Libraries and funded by a grant from the Andrew W. Mellon Foundation, established a task force charged with examining "new options for the collection and dissemination of scientific and technical information that could break an extraordinary information cost spiral while at the same time support innovative applications of information technology." (*Report of the Task Force . . .* 1994) Task Force members included scientists, university librarians, and university administrators as well as representatives from associations. The Task Force employed an analytical approach that described the system of scientific and scholarly communication by identifying functions, performance attributes, and system participants. The defined framework then provided a basis for formulating and evaluating new models for the management of scientific information.

The Task Force described three distinct models that it termed "classical," "modernized, and "emergent." *Classical* identifies the traditional, print-based model as described by Garvey and Griffith. The *modernized* model, defined in three variants, employs a mix of paper and electronic communications; it compares to Lynch's characterization in which new technology permits a more cost-effective and/or efficient means of continuing an established practice. The *emergent* model is related to Lynch's transformed approach where technology changes processes in fundamental ways.

Also, the Task Force identified a number of innovative initiatives such as the Human Genome Project (described fully in Chapter 5 and used in this chapter as examples in the models described). The Task Force Report provides a thorough and thoughtful discussion that recognizes the complexity of the communication system; its analysis has contributed significantly to this chapter.

Another perspective is provided by Karen Drabenstott (1994) in an analytical review of literature on libraries of the future. Her focus is on digital libraries, but she relates these developments to the entire system of scholarly communication. The issues she identifies and the problems she describes are similar to those discussed here. Her review is a balanced presentation that provides insights and direction to those who wish to play a part in disseminating, organizing, or managing information in the coming decades.

New Models for Scientific Communication

Computer-based communication, identified earlier as a powerful change agent in the system of scientific communication, is not, however, the only force exerting pressure on established modes of operation. New research directions that will also exert influence include globalization of research, increasing interdisciplinarity, and a focus on mission-oriented projects. Derek Price (1963) popularized use of the term "Big Science" to capture the changes that occurred gradually during the middle of this century. Big Science is seen not merely as a change in scale, i.e., the numbers of scientists engaged in research; more accurately, it is characterized by increased collaboration resulting in more multi-author papers and increasingly costly experiments. Now, more than thirty years later, the changes Price described have accelerated. Perhaps the present might be considered an era of "Bigger Science." Price wrote of "multi-author papers" with three and four authors; in 1994 over 400 of the papers indexed in the *Science Citation Index Source Index* had more than 50 authors! (Pendlebury, quoted in *New Scientist* 1995) Some papers in high energy physics have literally hundreds of authors, although these are outliers at present. "Bigger Science" thus is a fundamentally different model than that of small teams, which may be made up of graduate students working closely with a faculty member, and possibly a few post-doctoral fellows, within a single academic department in a university; or small groups of scientists and technicians based in a government laboratory or research institute.

The number of scientists involved in an experiment attests to its scale and complexity. Such projects may be global efforts as in high energy physics or the space sciences where large-scale installations, such as particle accelerators and astronomical observatories, collect massive computer-stored sets of data in numerical and image formats. Other projects are mission-oriented and directed toward solution of societal problems related to the environment or public health. Teams working on these problems are drawn from various disciplines in order to bring together all needed knowledge and skills.

20

"Bigger Science" may be well served by new modes of communication; traditional communication channels may be too slow or otherwise limiting to those engaged in large-scale interdisciplinary efforts. Examples of initiatives described in this and following chapters suggest that these very active research specializations may use technology in new ways and be responsible for innovations that will later be adopted by other disciplines. The discussion to follow draws on both experimental and operational developments that offer promise of changing the current system of scientific communication. Several models are presented that represent a series of evolving systems, beginning with a modernized version of the Garvey/Griffith model and moving toward a transformed model for scientific communication. These models are speculative but soundly based in the reality of numerous initiatives currently underway. The future may not match exactly what is predicted here, but it will certainly contain some elements of these models.

A Modernized Garvey/Griffith Model

Aspects of this model were described over fifteen years ago by Lancaster (1978) in his book on "paperless information systems" even before the Internet was as fully developed and as far-reaching as we know it today. By recognizing that every element of the traditional model has been affected by information technology, a modernized model can be outlined. Although electronic-based, it retains the key feature of the well-established system by building on the peer-reviewed scientific journal as the basic unit of distribution. [Figure 2-2]

In fact, this model represents reality for some specializations, although, at present, it co-exists with the paper-based communication system. The following examples illustrate this updated paradigm.

Informal communication among scientists has changed significantly through the already well-established use of electronic mail and listserves. A researcher no longer finds it necessary to telephone or travel to discuss research with a colleague. Conversations between individuals separated by large distances can occur at times convenient to each participant. Larger groups can discuss common interests on listserves that could be said to support an "electronic invisible college," whether open or closed to all but approved subscribers. Estimates of the number of such discussion groups vary widely; numbers in the range 12,000 to 15,000 have been reported. Some 1800 have been identified by a team led by Diane Kovacs at Kent State University Libraries as being of particular interest to scholars, researchers, and students. Users of the Internet can attest to the existence of a listserve for virtually any specialized interest. (King, Kovacs, et al 1994)

21

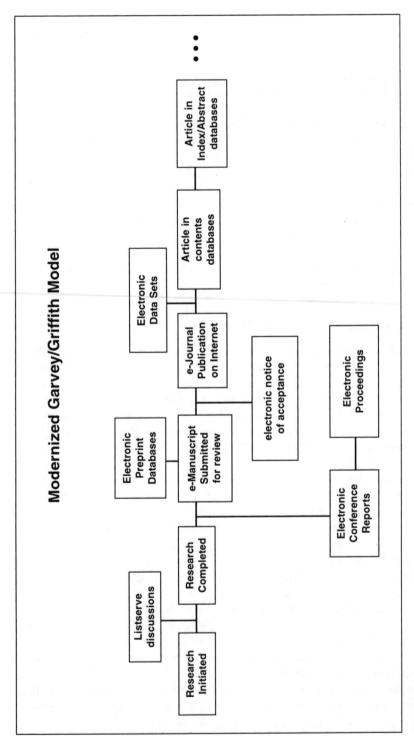

Figure 2-2

As a research project reaches completion, technology offers opportunities to share findings with colleagues through electronic conferences such as CHEM-CONF and ECTOC-1 or through electronic preprint distribution. Preprint databases have already become widely accepted in high energy physics. (See Chapter 4 for a more detailed discussion.) Most authors now compose manuscripts using word processors and may also submit articles to journal editors either across the Internet through a file transfer process or by mailing a disk with a copy of the manuscript. Reviewers can submit comments electronically as well, although some express concerns about issues of privacy in an electronic medium. Use of technology offers potential to shorten significantly many of the delays characteristic of a paper-based process while retaining the value-added features of peer review and editing that continue to make the refereed scientific journal an important component of the academic reward structure.

Electronic Journals

Electronic journals are increasing in number. The Association of Research Libraries annually publishes the *Directory of Electronic Journals, Newsletters, and Academic Discussion Lists*; the directory strives for comprehensive coverage of existing journals and newsletters. Compilers of the directory search Internet sites and solicit contributions of new entries. Ann Okerson, writing in the 1994 directory, identified over 440 electronic journals and newsletters, nearly double the number reported the previous year (ibid). Some electronic journals exist in purely computer-based format; others are equivalents of paper publications. Peer-reviewed titles are increasing in number with 74 reported by Okerson in 1994. The *Online Journal of Current Clinical Trials* and the *Chicago Journal of Theoretical Computer Science* are peer-reviewed journals that have no paper counterparts. They are distributed to subscribers on workstations that run proprietary software, and the journal articles are displayed in a format closely resembling pages in a printed publication. Tabular data and images are high quality and incorporated into text. Hypertext links in the articles make connections with other related information such as cited articles. These journals demonstrate how technology can support a familiar mode of distribution of information by speeding processing and dissemination, and add value by linking related information.

Much interest exists in electronic formats among commercial and scientific society publishers of print journals. Many publishers are experimenting with new modes of delivery of information, often working in collaboration with institutions such as university libraries. The CORE (Chemistry Online Retrieval Experiment) project, based at Cornell University, is one of the longest term studies. CORE is being carried out in collaboration with the American

Chemical Society, Bellcore, the Chemical Abstracts Service and OCLC. A digital library of chemistry literature was created along with a retrieval and delivery system providing access to faculty and students through their desktop computers. Multiple interfaces were designed utilizing both images and marked-up text; users can select from among a variety of ways to display, search, and navigate the database. CORE recently concluded seven years of data gathering and analysis that will soon be reported in the professional literature (Olson 1995). Among the many experiments underway, of which the CORE Project is only one, those that will prove not only attractive to users but economically viable are yet to be determined. At present the variety is staggering, and Okerson claims that "there has not been such a fruitfully chaotic time since possibly the invention of print or at least the origin of the journal 300 years ago." (King, Kovacs, et al 1994)

The modernized Garvey/Griffith model retains the basic elements of a traditional paper-based system by building on peer-reviewed journals as the unit of distribution for research. Moving from a paper to electronic medium accelerates the communication process and disseminates research findings more rapidly at all stages in the stream of communication.

Perhaps as significant is another feature of networked communication—its potential for opening the process to individuals previously excluded. If listserves are open to all (although it must be recognized that some are closed), scientists can participate in discussions whatever their institutional affiliation or geographic location. Similarly, electronic conferences overcome geographic and financial constraints, although participants must have suitable equipment and connectivity. Preprint databases, as yet unpublished in journals, provide access to research to far wider groups than were privy to such information when paper copies were mailed to lists of colleagues. This model represents a modernized system with enhanced opportunities for faster and wider communication through networked technology; it is not a transformed paradigm.

The No-Journal Model

This next model steps further away from the original Garvey/Griffith model by eliminating the journal as the distribution unit. It continues to recognize the role of peer review in validating scientific research, but it is built on the "article" or research report as the unit of distribution. [Figure 2-3]

Journals devoted to scientific studies have existed for over 300 years, having developed within early scientific societies to report to the membership at large on research carried out by individual members. Early journals were multidisciplinary, reflecting the broad scientific interests of their sponsoring groups. As scientific specializations emerged and became part of university curricula in the

24

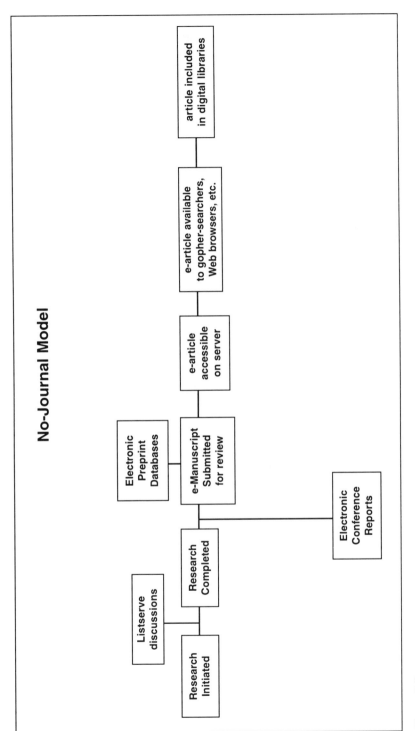

No-Journal Model

article included in digital libraries

e-article available to gopher-searchers, Web browsers, etc.

e-article accessible on server

e-Manuscript Submitted for review

Electronic Preprint Databases

Research Completed

Listserve discussions

Research Initiated

Electronic Conference Reports

Figure 2-3

late nineteenth century, the content of journals became increasingly specialized, a trend that continues to the present. In a communication system dependent on print on paper, packaging of some number of articles into a journal issue provides convenience for users as well as economies of scale in both production and distribution. An electronic distribution system offers the advantages of more frequent distribution, and in smaller basic units.

The "No-Journal" model outlines how networked communication might support distribution of electronic articles. Like the modernized model presented in the preceding section, elements of this scenario can be found in speculative writings and in initiatives underway. Charles Bailey (1994) summarizes models of scholarly electronic publishing that might be used, particularly by noncommercial publishers. The "Acquisition-on-Demand" (also known as "just in time") model places scholarly articles on network file servers to be retrieved as needed, for a fee. Quality filters, such as the number of previous uses, would aid users in judging the importance of papers, but otherwise, peer review is not mentioned in the model.

Bailey also describes a "Scholarly Communication System" first articulated by Sharon Rogers and Charlene Hurt that suggests how peer review might function in a system with the article as the distribution unit. Their proposal describes a paper that would initially be published for a six-month period, during which time use of the paper and citations to it would be monitored; in addition, comments would be appended by readers. After this period, the author could revise the paper for evaluation by a review board, which would then assign the paper to one of several categories, including one equivalent to rejection by a traditional journal. The system is envisioned as comprehensive and interdisciplinary and would be funded employing a mix of foundation, government, and university sources.

Another, related approach also described by Bailey is distributed, rather than centralized: a "Discipline-Specific Literature Base Model." A number of professional associations appear to be planning initiatives to allow testing of this type of distribution system, although, initially, they seem to be organizing information as "journals" even though dissemination of separate articles may be a possible outcome. In these models questions of copyright need to be addressed and licensing agreements are likely to be part of such systems, especially when associations are involved in building a database.

Ronald LaPorte and colleagues (1995) describe a Global Health Network built on a worldwide health information server that would provide access to biomedical articles through a wide area information system (WAIS). The group proposes a role in the system for peer review and describes how it might operate. The authors argue that use of the term "article" is inappropriate because hypertext links in the research communications will permit readers to move through multiple documents. The network is described as a dynamic system where communications can be continually changed and updated to reflect new discoveries, and the group is optimistic that development of such a system would address excessive delays in disseminating research results while simultaneously reducing expenses and greatly expanding access. The discussion, however, does not address the question of the source of financial support for a Global Health Network.

The Unvetted Model

Eliminating the component of peer review in a communication system is a transformation that some scientists have proposed. The system that might result we refer to here as an "Unvetted" model. [Figure 2-4]

Peer review has been criticized as favoring those whose research falls within established scientific paradigms and who are affiliated with the most prestigious institutions. Critics maintain that those whose findings question current theories and those whose work spans discipline boundaries often experience difficulties in getting articles accepted for publication. In a paper-based communication system, it proves difficult to design alternative forms of publication that do not suffer from perceived lack of standing and that are economically viable for all authors. A network-based communication system offers several approaches to publication of articles and utilizes resources that are now available to large numbers of scientists.

The electronic preprint databases that have become the accepted mode of distribution in high energy physics and a few other specializations represent one approach to a free and open communication system. (Use of the term "free" suggests lack of barriers rather than no-cost, as such systems depend heavily on institutional infrastructures to support large computers attached to the Internet.) The physics databases are discussed more fully in Chapter 4, but the centralized approach they employ is described generally here to suggest how articles might be made available without the constraints of peer review.

In the traditional paper-based system, sharing of preprints was an established practice, although distribution was generally restricted to those the author considered to be peer institutions and invisible college colleagues. This form of communication was regarded as important enough that many research libraries maintained preprint files to organize the collections that were heavily used.

An electronic approach also relies on institutional infrastructure but uses computer resources rather than library space and organizational skills. A host organization provides storage space on a server in which authors deposit papers representing completed research. Links to other papers as well as to data

27

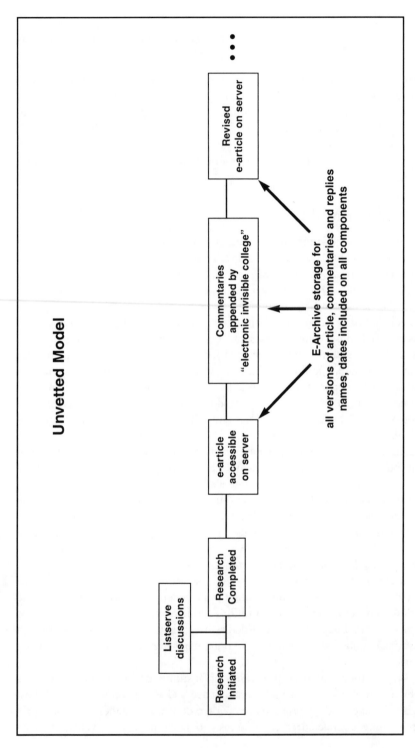

Unvetted Model

Research Initiated

Listserve discussions

Research Completed

e-article accessible on server

Commentaries appended by "electronic invisible college"

Revised e-article on server

E-Archive storage for all versions of article, commentaries and replies names, dates included on all components

Figure 2-4

archives, images and other information may be incorporated. The host computer maintains a list of all the papers and may notify interested individuals, through a listserve, of the addition of new papers to the database. The papers are available on the server for readers; they may also be transferred to other machines using standard file transfer protocols.

The World Wide Web and interfaces such as Mosaic and Netscape that permit browsing of the Web offer a graphically rich format that some scientists are beginning to use to disseminate their research. Unlike the preprint databases just described, this is a more distributed system where each author "publishes" completed papers on a "home page," a Web document residing on a particular user's computer but accessible to anyone connected to the Internet. Currently, users browse the Web by activating software agents such as Lycos and WebCrawler that were developed to search words in titles, headers and even text of Web documents; some search engines search indexes and directories as well. The role proposed for "knowbots" that would collect and filter network data is increasingly important as the amount of material proliferates.

The distributed approach of the Web model trades rapid, open access for qualities added by both editors and reviewers. Rather than collecting articles on a specialization, as in the preprint database, the communication system is dispersed and searched with individualized criteria for each user. Some observers consider it to be a "vanity press" model, a point of view that reflects the perceived values added by peer review and editorial processes.

The Collaboratory Model

Furthest removed from the original Garvey/Griffith model, and representing a genuinely transformed communication system, is the "Collaboratory" model. [Figure 2-5]

The term "collaboratory" was coined in a National Research Council report and melds the notion of collaboration with that of laboratory to convey an image of a worldwide network of computers supporting a global research community (Wulf 1993). Scientists in a collaboratory exchange data, share computer power, and consult digital library resources, interacting across great distances as easily as if they were sharing a physical facility. The collaboratory concept is particularly applicable to those projects of "Bigger Science" requiring large-scale instrumentation such as observatories and space satellites or enormous shared databanks such as the one under construction in the Human Genome Project. (Subsequent chapters provide more details on work in the space sciences and the Genome Project.) This discussion examines a communication system that might prevail in the collaboratory environment, where data items are the units of information exchange.

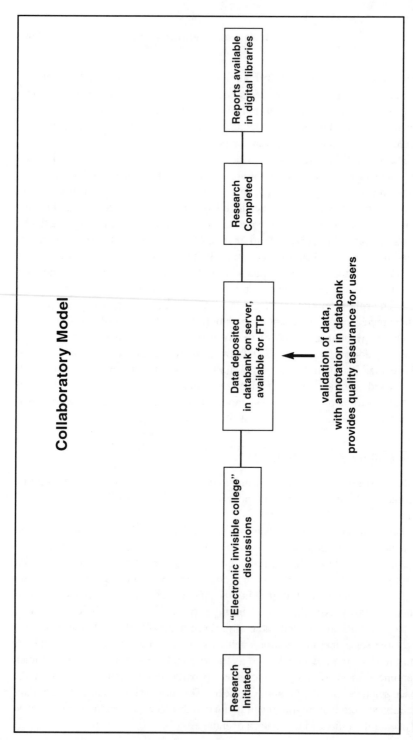

Collaboratory Model

| Research Initiated | → | "Electronic invisible college" discussions | → | Data deposited in databank on server, available for FTP | → | Research Completed | → | Reports available in digital libraries |

validation of data, with annotation in databank provides quality assurance for users

Figure 2-5

National Science Foundation (NSF) support has made possible the development of prototype software to facilitate collaboration among widely scattered research groups in space physics, known collectively as the Upper Atmospheric Research Collaboratory (Watkins 1994). Participants in the project study such phenomena as the interaction of solar wind and the Earth's magnetic field by drawing on data collected by remote instrumentation. With collaboratory conferencing software running on NeXT workstations, team members can share and manipulate data in real time.

Altogether the NSF has funded twelve collaboratory projects. The grants not only advance the basic research underway in each group but also provide behavioral scientists with an opportunity to study how scientists communicate in a collaboratory environment. For example, psychologists have collected baseline data on communication practices among a group of physicists and are now engaged in monitoring how participants use their workstations to communicate. Some of their earliest findings describe how readily the physicists assimilated communication technology into work habits, and the findings also identify potential benefits for graduate students involved in the research that arose from the enhanced opportunities for communication (ibid).

In the collaboratory environment, data from many participants are entered into shared databases and issues of accuracy surface early in the process of building those databases. All the collaboratory efforts described in following chapters—the Human Genome Project, which is mapping human gene sequences; space scientists building databases of astronomical measurements; and the Worm Community System for biologists studying a species of nematode worm (Schatz 1992)—have devised techniques to assure data quality. In effect, they build in a sort of peer review at an early stage of the communication process. The prototype collaboratories are also experimenting with approaches to sharing commentaries on research and accommodating revisions of previous electronic publications; some are even puzzling over issues relating to archival records that document the emergence and development of ideas.

Future Directions

This chapter has surveyed scientific communication and speculated on how information technologies and global computer networks might transform exchange of scientific information. The Garvey/Griffith model of scientific communication was outlined and four models that might evolve from that traditional model have been described. For each model presented, examples of current or proposed initiatives have illustrated how a particular aspect of the model might operate. A number of complex issues of economic, political, and

31

sociological nature have been identified but not fully explored: we will revisit those issues in the final chapter.

We should also emphasize here that any changes will likely be evolutionary rather than revolutionary, and that new communication systems will likely co-exist alongside the old model for some time to come. At the same time, the developments that are catalyzing significant change are themselves dynamic. What seems obvious now as a future direction may ultimately prove only a brief perturbation and the real future something not yet imagined. The question for which only time will provide an answer is: How will technology alter the basic structures that support scientific communication and lead to a system substantially different from one derived before computer use was widespread?

References

Arthur, C. 1995. And the Net total is . . . *New Scientist* no. 1977 (May 13):29-31.

Bailey, C. W., Jr. 1994 Scholarly electronic publishing on the Internet, the NREN, and the NII: Charting possible futures. *Serials Review* 20, no. 3:7-16.

Bush, V. 1945. As we may think. *Atlantic Monthly* 176 (July): 101-108.

Communications of the ACM. 1995. Special issue on digital libraries 38 (April), no. 4.

Crane, D. 1972. *Invisible Colleges: Diffusion of Knowledge in Scientific Communities.* Chicago: University of Chicago Press.

Cronin, B. 1982. Progress in Documentation. Invisible Colleges and Information Transfer: A Review and Commentary with Particular Reference to the Social Sciences. *Journal of Documentation* 38, no. 3 (September): 212-236.

DeLoughry, T. J. 1994. For the community of scholars, "being connected" takes on a whole new meaning. *The Chronicle of Higher Education* A26 (November 2).

Drabenstott, K. M. 1994. *Analytical Review of the Library of the Future.* Washington, DC: Council on Library Resources.

Garvey, W.D. 1979. *Communication: The Essence of Science.* Elmsford, NY: Pergamon Press.

Garvey, W. D. & B. C. Griffith. 1972. Communication and information processing within scientific disciplines: Empirical findings for psychology. *Information Storage and Retrieval* 8:123-126.

King, D. W. and N. K. Roderer. 1982. Communication in physics—the use of journals. *Physics Today* 35, no. 10 (October):43-47.

King, L. A., D. Kovacs, and the Directory Team. 1994. *Directory of Electronic Journals, Newsletters and Academic Discussion Lists.* 4th edition. Ann Okerson, ed. Washington, DC: Association of Research Libraries.

Krieger, J. H. 1995. Organic chemistry conference on Internet sets new pace. *Chemical & Engineering News* 73, no. 34 (August 21):35-38.

Lancaster, F.W. 1978. *Toward Paperless Information Systems.* London: Academic Press.

LaPorte, R. E., E. Marler, S. Akazawa, F. Sauer, C. Gamboa, C. Shenton, C. Glosser, A. Villasenor, and M. Maclure. 1995. The death of biomedical journals. *British Medical Journal*, 310 (May 27):1387-90.

Lederberg, J. 1993. Communication as the root of scientific progress. *The Scientist* February 8, 1993:10-14.

Licklider, J.C.R. 1965 *Libraries of the Future*. Cambridge, MA: M.I.T. Press.

Lynch, C. 1993. The transformation of scholarly communication and the role of the library in the age of networked communication. *Serials Librarian* 23, no. 3:5-20.

O'Haver, T. C. 1994. CHEMCONF: An experiment in international on-line conferencing. Paper presented at the Fourth TriSociety Symposium on Chemical Information held at the Annual Meeting of the American Society for Information Science, Alexandria, VA, October 17, 1994. An article based on the presentation is in the *Journal of the American Society for Information Science* 46, no. 8 (September 1995): 611-613.

Olson, J. K. 1995. Presentation at the American Library Association Annual Meeting, Association of College and Research Libraries, Science and Technology Section, Forum on Science and Technology Library Research, June 25.

Oseman, Robert. 1989. *Conferences and their Literature: A Question of Value*. London: The Library Association.

Pendlebury, D. 1995. Quoted in *New Scientist* 29 (April 29).

Piternick, A. B. 1989. Attempts to find alternatives to the scientific journal: A brief review. *Journal of Academic Librarianship* 15, no. 5:260-266.

Price, D. J. de Solla. 1963. *Little Science, Big Science*. New York: Columbia University Press.

"Report of the Task Force on a National Strategy for Managing Scientific & Technical Information." 1994. Submitted to the Steering Committee of the Research Libraries Project, Association of American Universities. Washington, DC, April 4.

Rider, F. 1944. *The Scholar and the Future of the Research Library*. New York: Hadham Press.

Schatz, B. R. 1992. Building an Electronic Community System. *Journal of Management Information Systems* 8, no. 3:87-107.

Watkins, B. T. 1994. A far-flung collaboration by scientists. *The Chronicle of Higher Education* A15 (June 8).

Wilson, D. L. 1991. Testing time for electronic journals. *The Chronicle of Higher Education*. A22-A24 (September 11).

Wulf, W. A. 1993. The collaboratory opportunity. *Science* 261 (August 13):854-855.

Chapter Three

The Human Genome Project

Ann C. Weller

The human genome project . . . is the source book for biomedical research in the 21st century and beyond.

—*Mark Guyer and Francis Collins (1993)*

The decision to undertake the mapping and sequencing of the human genome has been likened to President John F. Kennedy's 1961 decision to fund a project to put an American on the moon. Both goals took vision and the inevitable gamble of pursuing a goal of unknown value or uncertain success. Although the total funding for the human genome project (HGP) will be less than that of the lunar exploration, the consequences may prove much more important. Sometime in 2005, the international effort to construct the physical maps, the genetic maps, and the DNA sequencing for the approximately 100,000 human genes on the twenty-three pairs of human chromosomes will be completed. These data will provide an invaluable resource for localizing and isolating any site of interest on the human genome (Collins and Galas 1993).

Proponents of the project cite how mapping the human genome will advance medical sciences and assist health professionals to understand and treat the more than 4,000 genetic diseases that now afflict the human race. Nonetheless, the actual extent to which genetics can play an effective role in disease treatment and prevention is debatable. Holtzman (1989) gave three reasons for caution when using genetic information: "(1) the past use of genetics in public and corporate policy; (2) the current trend toward commercialization in biotechnology; and (3) the disparity between diagnosis and effective intervention." Holtzman published his cautionary book the year after the U.S. Congress approved initial funding for the HGP. His work stands as a solid introduction to the scientific, political, social, ethical, and confidentiality issues the HGP has raised.

The size and complexity of this project, with its international scope and reliance on computers and telecommunications, challenges the traditional model of scientific communication, including the role of invisible colleges. This chapter describes the HGP's progression from a paper-based to a computer-based project. The scientific communications issues covered in this chapter consist of the following sections: historical perspective; development and status of the HGP; databases containing genetic information; electronic data publishing; access issues and policies; changes in communication among scientists; model of invisible colleges; and unresolved issues.

Historical Perspective

Over a century ago, in 1866, Mendel published results of his experiments with fertilizing varieties of the garden pea. He recorded the frequency of a number of plant characteristics such as tallness or dwarfishness, and the presence or absence of colors in the blossoms, among others. From his analysis, he introduced a new theory of inheritance. Not until almost 40 years later, in 1900, three researchers—Hugo de Vries in Holland, Carl Erich Correns in Germany, and Erich Tschermak von Seysenegg in Austria—independently conducted similar research and reached similar conclusions. Each one duly credited Mendel. Today, ready access to information would, in theory, prevent such an unfortunate delay in communicating a major new scientific theory.

All scientific discoveries, of course, build on previous ones. Credit for one of the most highly publicized genetics-related discovery goes to James Watson and Francis Crick (1953) when they "suggested" (their words) in a letter to the editor published by *Nature* that deoxyribonucleic acid (DNA) had a double helical structure. As the information in Table 3.1 (page 40) shows, Watson and Crick's discovery was actually one in a long line of scientific advances that eventually resulted in discovering the physical and chemical structure of DNA.

In 1965 Margaret Dayhoff and her colleagues published the first compilation of sequencing information in the *Atlas of Protein Sequence and Structure* (Dayhoff et al. 1965). Their purpose was to "collect between a single pair of covers as many as possible of the known protein and nucleotide sequences." (Dayhoff et al. 1967-68, p. vi). The first *Atlas* comprised fewer than 50 sequences of 30 links or more. The second one doubled the number of sequences to 100 (ibid. p. viii). The diagrams of these sequences were added as fourteen loose-leaf plates to the 1967 edition. This early work is important for a number of reasons:

- all protein sequence information was codified and disseminated to anyone who asked for it;

- sequence information was reviewed and corrected—up to 15 percent of the data were revised as a result of the review process;
- computers were used to help with this task; and
- information on 1,400 authors with an accompanying file of preliminary reports, preprints, and corrections to previous work was maintained.

In effect, Dayhoff and her colleagues had established a network of 1,400 genetic specialists. By the time the last edition of the *Atlas* was completed (it was published irregularly until 1978), it had become quite unwieldy and probably impossible to continue as a printed work. Consequently, a clear need existed to build a fully supported database of genetic information (Smith 1990).

Development and Status of the HGP

Throughout the 1980s tremendous progress was made in the storage capacities of computers, the availability and ease of using personal computers, and the increased access to vast amounts of information available via telecommunications.

In 1979 a group of scientists meeting at Rockefeller University concluded that there was a need for a government-supported nucleic acid sequence database (Smith 1990). The group's recommendations and a series of meetings held through 1988 led to the Congress's decision in 1988 to support the HGP (Table 3.2). The cooperation between the Department of Energy (DOE) Office of Health and Environmental Research (OHER) and the National Institutes of Health (NIH) also played a key role in facilitating initial funding. The decision to fund the HGP jointly from two major U.S. institutions, DOE and NIH, was not without controversy. Koshland (1987), editor of *Science*, argued that the two agencies complemented each other: DOE had sufficient experience with large projects and NIH had funded most of the scientists already involved with genetic research. In 1988, DOE and NIH signed a memorandum of understanding for interagency coordination of the genome project (Roberts 1988). Congress funded $17.2 million for the first year and Watson was named project director at NIH (Watson 1990).

The first five-year goals (1991-95) for the HGP were revised in 1993 in light of the progress made the first few years; some early goals even ran ahead of schedule (Casey 1994). Specific scientific goals for the U. S. component of the project for 1994-1998 cover nine major areas:

- Mapping and sequencing the human genome, including specific goals for genetic maps, physical maps, and DNA sequencing;
- Gene identification, incorporating the known genes into the physical map;

Table 3.1

Major Genetic Discoveries

Date	Event
1944	Oswald Avery and his colleagues open the door to the sequencing of the genetic code by reporting that the structure of genetic material consisted "principally, if not solely, of a highly polymerized viscous form of deoxyribonucleic acid" (Avery, MacLeod, and McCarty 1944, p. 156). Scientists had previously assumed that genetic material was a protein.
1950	Linus Pauling and Robert Corey (1950) described, in a letter to the editor of the Journal of the American Chemical Society the use of x-ray crystallography in identifying the helical structure of a protein.
1950	Erwin Chargaff (1950) observed that within DNA the concentration of thymine equaled that of adenine and the guanine that of cytosine.
1953	Rosalind Franklin and R. Gosling (1953) were able to deduce that DNA consisted of two interconverted structures.
1953	Watson and Crick (1953) suggested the double helical shape of DNA.
1956	Fred Sanger (1956) developed a method for directly determining the amino acid sequence of protein.
1973	Cohen, Chang, Boyer, and Helling (1973) began the recombinant DNA revolution with the construction in a test tube of a biologically functional DNA molecule that combined genetic information from two entirely different organisms.
1977	Allan Maxam and Walter Gilbert (1977), and Sanger et al. (1978) developed a method determining base sequencing techniques, greatly speeding the process of decoding DNA segments.
1978	V. Reddy et al. (1978) and W. Friers et al. (1978) completed the first DNA sequencing on small DNA viruses using the techniques of Maxam and Gilbert, and Sanger.

Table 3.2

Chronology of Major Events Leading to Funding the HGP

Date	Event
1979	A workshop sponsored by the NSF at Rockefeller University concluded that a fully supported international nucleic acid database was needed (Smith, 1990).
1980	NIH organized a "nucleic acid sequence data bank workshop" which made long- and short-term recommendations on the nature of a U. S. sponsored nucleic acid database (Smith, 1990).
1984	OHER, DOE and the International Commission for Protection Against Environmental Mutagens and Carcinogens co-sponsored a meeting to consider whether new methods could permit detection of DNA mutations, especially in the children and survivors of the Hiroshima and Nagasaki bombings Attendees recognized the importance of sequencing the human genome (Cook-Deegan, 1989).
1985	Sinsheimer (1989) organized a workshop at the University of California at Santa Cruz of leading molecular biologists who developed the first serious proposal to begin sequencing the human genome.
1986	A meeting in Santa Fe, New Mexico sponsored by the DOE assessed the desirability of implementing a project to map the human genome (Watson 1990).
1986	A meeting at the Cold Spring Harbor Laboratory initiated by DOE debated the usefulness of spending NSF funds for sequencing data which potentially could drain funds needed for basic research (Watson 1990).
1987	Report on the Human Genome Initiative was prepared by the Health and Environment Research Advisory Committee (HERAC) of the DOE (Department of Health and Human Services 1990).
1988	A report from the Office of Technology Assessment (1988), Mapping Our Genes . . . , summarized scientific developments and outlined a number of options as to how the United States could undertake a project of this magnitude.
1988	DOE and NIH signed a memorandum of understanding for interagency coordination on the genome project and estimated that the effort would cost $3 billion and take until 2005 to complete (Roberts 1988).
1988	Congress funded $17.2 million for the first year of the genome project and Watson was named director of the project at the NIH (Watson 1990).
1989	The National Center for Human Genome Research (NCHGR) (1995) was established to head NIH's role in the HGP, and became an independent funding unit within NIH with the authority to award grants. NCHGR is the U. S. link with HUGO to support international cooperation.

- Technology development, expanding support of innovative approaches;
- Modeling organisms, e.g., mapping the worm (C. Elegans), the mouse, and E. coli;
- Investigating the ethical, legal, and social implications;
- Training, focusing on interdisciplinary issues;
- Technology transfer, targeting primarily the transfer of data and information to industry;
- Informatics, including the creation, development, and operation of databases; the consolidation, distribution, and development of software; the development of tools for comparing; and the interpretation of genome information;
- Outreach goals, including cooperation with distribution centers, the incorporation of the move to the private sector, and the sharing of information within six months (Collins and Galas 1993).

The two final goals (informatics and outreach) focus on scientific communication—primary areas for discussion in this chapter. Project planners realized from the beginning that the informatics concerns of access, standards, storage, and organization of such large amounts of data, needed to be addressed. The informatics portion of the project has maintained a substantial portion of the funding; e.g., $12.7 million, or 20.3 percent of the total $62.4 million sponsored by DOE for fiscal year 1993 (Department of Energy, Office of Energy Research 1994).

The National Center for Biotechnology Information (NCBI) was established in 1988 to oversee the informatics components of the HGP, including database projects, and to provide the technical support needed to map the entire human genome. The National Library of Medicine (NLM) seemed the logical place to house NCBI, where it has remained, because NLM operates under NIH and played a leadership role in using innovative technologies and managing databases. NCBI works in conjunction with NIH's genome research organization, the National Center for Human Genome Research (NCHGR).

NCBI has a fourfold mandate (Denson et al. 1990):

- Create automated systems for storing and analyzing knowledge about molecular biology, biochemistry, and genetics;
- Perform research into advanced methods of computer-based information processing for analyzing the structure and function of biologically important molecules;
- Facilitate the use of databases and software by biotechnology researchers and medical personnel; and
- Coordinate efforts to gather biotechnology information worldwide.

40

Databases Containing Genetic Information

Databases have evolved to keep pace with the increasingly complex nature and growing volume of information that exist today in the HGP (Table 3.3). Human genome information is available in three types of databases: bibliographic, sequences, and maps.

The major bibliographic databases that contain genetic information consist of the following:

- AGRICOLA, produced by the National Library of Agriculture;
- BIOSIS, the online version of *Biological Abstracts*;
- CAS Online, the online version of *Chemical Abstracts*;
- MEDLINE, NLM's online version of *Index Medicus*.

That these bibliographic databases are also the major scientific and medical databases illustrate the interdisciplinary nature of genetics.

Two types of sequencing databases exist: nucleic acid sequence databases and protein sequence databases. The four major databases that contain nucleic acid sequencing information comprise the following:

- GenBank
- European Molecular Biology Laboratory (EMBL)
- DNA Data Bank of Japan (DDBJ)
- GSDB, the National Center for Genome Research, Los Alamos National Laboratory (LANL).

Producers of all four databases cooperate in an international effort to make available all known DNA and RNA sequences, from any organism, not limited to the human genome. They have worked collaboratively since 1987 to collect and share nucleic acid sequence data. Each of these databases collects a portion of the total sequence data reported worldwide. The content and format of their records are nearly identical. All share information nightly via satellite, producing databases that are essentially identical, but in slightly different formats (Adamson and Casey 1994).

Major protein sequence databases consist of the following:

- PIR-International, which combines protein sequence data from three centers: (1) Protein Information Resource (PIR), produced by the National Biomedical Research Foundation (NBRF), Georgetown University, Washington, DC; (2) JIPIDS (Japan International Protein Information Database), Tokyo, Japan; and (3) MIPS (Martinsried Institute for Protein Sequence Data), Martinsried, Germany (Barker et al. 1993).

Table 3.3

Chronology of Database Development for the HGP

Date	Event
1965	Margaret Dayhoff and R. S. Ledley assemble the first major collection of genetic sequence information (Dayhoff et al. 1965).
1973	The First International Human Gene Mapping Workshop (HGM1) was held at Yale University; 75 attendees summarized 25 genes that had been mapped in humans (Ruddle and Kidd 1989).
1980	The European Molecular Biology Laboratory (EMBL) Data Library began to maintain and distribute a database of nucleotide sequences (Rice et al. 1993).
1982	National Institute of General Medical Sciences (NIGMS) awarded a subcontract to LANL to build a nucleic acid sequence database named GenBank (EMBL Data Library 1987).
1984	BIONET, funded by NIH, was established as a computer network giving microbiologists unlimited access to genetic databases for government and nonprofit groups. Databases included those managed by Intelligenetics, LANL, and EMBL in Heidelberg (Databases: . . . 1984).
1984	CEPH was established in 1984 in France to maintain a genotype database which is distributed regularly and sponsors the construction of linkage maps (HNIH/CEPH Collaborative Mapping Group 1992).
1986	The Protein Identification Resource (PIR) was established as an online resource produced by NBRF to provide information on the identification and analysis of protein sequences and coding sequences (George, Barker, and Hunt 1986). This database is the continuation of the *Atlas* began by Dayhoff (Barker, George, and Hunt, 1990). PIR-International also provides access to the nucleic acid sequencing databases.

Table 3.3 (continued)

1987   BIONET supported 489 laboratories with 16 telnet ports and 6 direct dial ports (Kristofferson 1987). Federal support ended in 1989 (Ezzell 1989).

1987   The last printed distribution of the GenBank database, published in collaboration with EMBL, occupied 8 volumes and contained 8823 entries representing 8,442,357 base pairs (Burks et al. 1992).

1988   NCBI was established by Congress to manage and distribute molecular biology information (Woodsmall & Benson 1993).

1988   Expansion of the PIR databases took place as PIR-International incorporated three databases—PIR, MIPS, and JIPIDS—to maintain a complete set of published protein sequences in one place (Barker et al. 1993).

1989   A common language for mapping and tagging the human genome is proposed (Olson et al. 1989).

1991   GenBank began accepting direct submissions to its electronic files (Pearson 1991).

1992   GenBank is updated nightly via satellite at ten sites world wide (Burks et al. 1992).

1992   NCBI became responsible for managing GenBank.

1994   Argonne National Laboratory announced it had developed a "super chip" capable of sequencing genes 1,000 times faster and 10 times more economically than previously done (Gorner 1994).

1994   Merck & Company funded Washington University to sequence gene data. All information is added to GenBank with no restrictions. Sequences are being added at the rate of about 1,000 per day (GenBank adding . . . 1995).

1995   Integration of genome and genome-related databases being developed with a category of software called "middleware" to facilitate the construction of federated databases (Five years . . . 1995).

- SWISS-PROT (annotated protein sequence database), which follows the format of, and is compatible with, the EMBL (Cameron 1988).
- GenInfo, a database that links sequence data to the patent literature and the plant genome project.

The two major chromosome mapping databases are the Genome Data Base (GDB) and the online Mendelian Inheritance in Man (OMIN) database. GDB provides the genetic linkage maps that show how genes and other identifiable markers are inherited, providing location, ordering, and distance information on human genetic markers. This information is linked to McKusick's *Mendelian Inheritance of Man*, the text-based catalog of inherited human traits and diseases (Pearson and Soll 1991; Brandt 1993).

Each of the three types of databases has a host of electronic addresses for submitting and updating data, asking questions, searching, ftp-ing files, locating reports and statistics, and examining linkage maps, to name a few. BankIt, released in 1995, gives researchers the opportunity to add sequence data to GenBank directly through the World Wide Web (WWW). BankIt was developed in conjunction with EMBL and DDBJ, and has complete instructions for submitting material on the WWW. BankIt can handle sequences of up to thirty thousand nucleotides. As of September, 1995, seven thousand GenBank entries had been submitted via BankIt (BankIt . . . 1995).

Electronic Data Publishing

Journal editors played a crucial role in the transformation of genetic information to an online environment. It is interesting to trace the changes in editorial policies for manuscripts submitted to some of the major journals that publish sequencing information. These changes have had a dramatic impact on the way databases are maintained and updated. The link between raw data and the journal literature has changed. Traditionally, authors provided information on the collection, analysis, and discussion of the data, but not the data themselves. With the HGP, all data items are subject to the scrutiny of the editorial review process. The editorial review process, submission requirements, and the methods of making corrections have been modernized and become more opened.

In 1983, the *Journal of Biological Chemistry*, in its "Instructions to Authors," stated that "in view of the increasing numbers of submitted papers describing nucleic acid sequences . . . data obtained by well-established techniques . . . will generally not be published and should not be submitted with the manuscript" (Instructions to Authors 1983). The next year, that journal's editors became the first to ask authors to submit nucleotide sequencing data directly to a database, either GenBank or EMBL (Instructions to Authors 1984).

In 1987 the executive editors of *Nucleic Acids Research* announced that, in an attempt to "cope with the expected flood of sequence data once automation becomes generally available" (Walker et al. 1987) they would require all authors to go directly to EMBL, not GenBank, in order to maintain the two databanks roughly at equal size. The editors anticipated that at some point in the future the option of submitting directly to GenBank would be available. Authors were required to obtain an accession number from EMBL prior to submission of a manuscript to *Nucleic Acids Research*. A detailed description of implementation and instructions for data submission to EMBL was published with that announcement (EMBL Data Library 1987). Reviewers were given access to the data online to check the accuracy of the authors' claims or statements. Authors could request that their data be held confidentially until the time of publication, and approximately 20 percent did so.

Nucleic Acids Research updated the instructions in 1988 with an announcement of a new two-way online communication feature: copies of the data submission form could be obtained via a file server from EMBL. In 1988, another important enhancement enabled EMBL to share its data with GenBank, DDBJ, PIR, MIPS, and JIPIDS. Editors of *Nucleic Acids Research* required submission of sequence data electronically and promised to give authors an accession number for their use within seven working days (Kahn and Hazledine 1988).

In a similar move, Igor Dawid, chair of the editorial board of the *Proceedings of the National Academy of Science* announced in 1989 that, as a leading outlet for the publication of sequence information, the *Proceedings* would require authors to submit their sequence data in machine-readable form to GenBank (Dawid 1989). An accompanying editorial by Burks and Tomlinson (1989) provided a detailed description of submission procedures. Beginning in 1989, GenBank was updated quarterly and data was made available approximately three to five months after submission. With this time frame for making data available, Burks and Tomlinson estimated that the sequence data would appear in GenBank about the same time that a manuscript was published in the *Proceedings*.

Similarly, in 1989, the editors of the *Journal of General Microbiology* also adopted a policy that requested authors to submit data directly to GenBank (Gilna, Tomlinson, and Burks 1989). The announcement included a list of sixteen additional journals whose editors required the submission of data to one of the HGP databases. DDBJ, EMBL, and GenBank divided journals so that any one journal needed to interact with only one of these database suppliers (Burks and Tomlinson 1989). One advantage of submitting data to a database prior to the publication of the manuscript is that the database staff can spot errors and alert the author. The problems can then be corrected prior to publication. The sequence accompanying the manuscript is labeled as "unpublished" prior to publication.

Cinkosky, et al (1991), called this form of publication "electronic data publishing," which, they said (p. 1274) "uses a highly structured, networked-based communication channel through which scientists can present their experimental results to others with a minimum of effort." They discussed the influence electronic data publishing has had on the editorial peer review process since data are added directly into a database. While the actual data have not gone through the traditional editorial peer review process, several potential sources of error introduction have been eliminated. In the traditional process of publishing articles, data are often transcribed, frequently more than once. Each transcription adds one more possibility for error introduction, especially in producing tables, figures, and charts for printed publications.

Even more important, perhaps, in the traditional editorial peer review process, reviewers usually do not have access to original data. Since 1983 at least, the "Instructions to Authors" section of the *Journal of Biological Chemistry* states that reviewers may access original data as submitted to one of the databases. Cinkosky et al. (p. 1276) argued that electronic data publishing actually increases the level of review that data receive before they are made available to the scientific community. In support of this claim, Rice, et al. (1993) have found that 90 percent of all data are directly submitted to EMBL, but that processing the remaining 10 percent is time consuming, error prone, and incomplete.

The innovation of adding data to databases at the time of manuscript submission was not without controversy. McGourty (1989), in a letter to *Nature*, addressed the fact that years of refinement often are necessary before a structure is ready to be entered into a database. She was concerned about rumors in the crystallography community, which claimed that researchers were selling coordinates to private companies instead of making them publicly available, and she argued that journals should play a role in the fast deposition of coordinates. Maddox (1989) opined that editors should not require authors to submit data to a databank. Cameron, Kahn, and Philipson (1989) from EMBL stated that the journal review process ensures the scientific integrity of what is published and asked the editors of *Nature* to require authors to submit the accession numbers received from EMBL during the publication process.

By 1995, approximately 100 journals regularly published sequence data. GenBank staff under NCBI routinely peruse the "Instructions to Authors" sections of these journals to make sure that they contain correct submission information. Because all information is now shared, restrictions no longer exist on which nucleic acid sequencing databases will accept submissions from which journals (Adamson and Casey 1994).

The evolution to electronic data publishing required cooperation between the database producers and journal editors. Initially, journal editors requested that authors obtain an accession number. Concerns similar to McGourty's about the privatization of information were addressed by the producers of GenBank, one of those databases. GenBank now requires that authors submit sequence information at the time of manuscript submission, if funds for their research came from any federal agency. This move assures that information is made public and not diverted to the private sector when the project depends on public sources for funding. Of course, enterprises that generate genetic data without federal funds are under no such obligation. Even some in the private sector, however, understand the advantage of making genetic information available through these databases. For example, Merck and Company announced in October, 1994 its intention of developing the human mRNA sequence as a public resource with access to the data unrestricted (GenBank adding . . . 1995).

A shift in the publication process began as the volume of sequence information increased. With electronic data publishing a new model of scientific communication has emerged. "[This] new model for scientific databases in which the community [is] directly . . . involved in the maintenance of databases" (Cinkosky et al. 1991, p. 1274) is a trend that will probably continue to expand and be applicable to more and more disciplines. One undisputable advantage of the move to electronic data publishing is the scientific community's direct access to genetic data.

The increasing internationalization of genome research has brought with it partnerships in Great Britain, France, Italy, and Japan, to name a few, all working together. The HGP differs from traditional research in that it creates a body of data, or "reference information" that will be used by a number of scientific disciplines (Yager, Nickerson, and Hood 1991).

The international character of the HGP has led to the establishment of a formalized review process through the Human Genome Organization (HUGO), a panel whose members represent many countries. HUGO was formed in 1991 and given the responsibility for reviewing data submitted to the GDB. HUGO has appointed reviewers for each chromosome to assure uniformity and correctness. HUGO also coordinates international efforts to assure work progresses without duplicating efforts (Pearson et al. 1991). Reviewers are recommended by peers and are responsible for validating data and providing guidance in moving the information to a public database. Their names are published in *Human Genome News* (Chromosome editors 1994).

HUGO's review procedure has altered the model of anonymity in the editorial review process; but it has not brought about an end to peer review for manuscripts with genetic information because the manuscripts themselves still undergo the scrutiny of anonymous editorial peer review. The manuscripts contain information on methodology, results, discussion, and conclusions that carry essential explanatory information. The result is a continuation of the move toward open access to data and more communication among scientists.

The checks for accuracy of genetic data take place after data have been entered. By 1991, a delay of about one year occurred from the time most sequence information was added until it was available in the public access databases.

> This interval seems long enough for the originating laboratory to ascertain its possible biological or commercial value and obtain competitive scientific advantage from the effort put into its acquisition but it is not so long as to incur the scorn of the scientific community or the wrath of the funding agencies (Pearson and Soll 1991, p. 37).

At times the review process changes data, and a problem can result if the data are not corrected in the electronic files. The latest release of the GDB software (autumn 1995) has a feature that permits anyone to make an annotation to the data, pointing out any inconsistencies or problems. This capability provides a method for researchers to enhance the work of other scientists and to move toward a community-owned database.

As shown in information on chromosome 12 from the Web site at Yale University (Figure 3-1), which was obtained with no special password or software (other than network software), information is available to anyone to examine, critique, or comment upon.

Access Policies and Issues

A recently published 350-page guide by Martin Bishop (1994), detailing information on genome computing, dramatically illustrates just how complicated genetic databases can be. This guide is an impressive documentation on how a variety of specialists—including geneticists, molecular biologists, informatics specialists, computer scientists, mathematicians, and statisticians—have worked together and expanded the knowledge base of genetic information.

Martin, Primich, and Riley (1994) listed twenty-two databases that they believe librarians should understand in order to handle genetic reference questions. A chart of these databases shows how each was available: gopher, client/server, CD-ROM, fee-based, telnet, or a database on a disk. Although several were available by more than one access point, none can be accessed by each option. Rice, et al. (1993), in a similar table, listed thirty-three databases distributed by EMBL. The authors also included information on access options

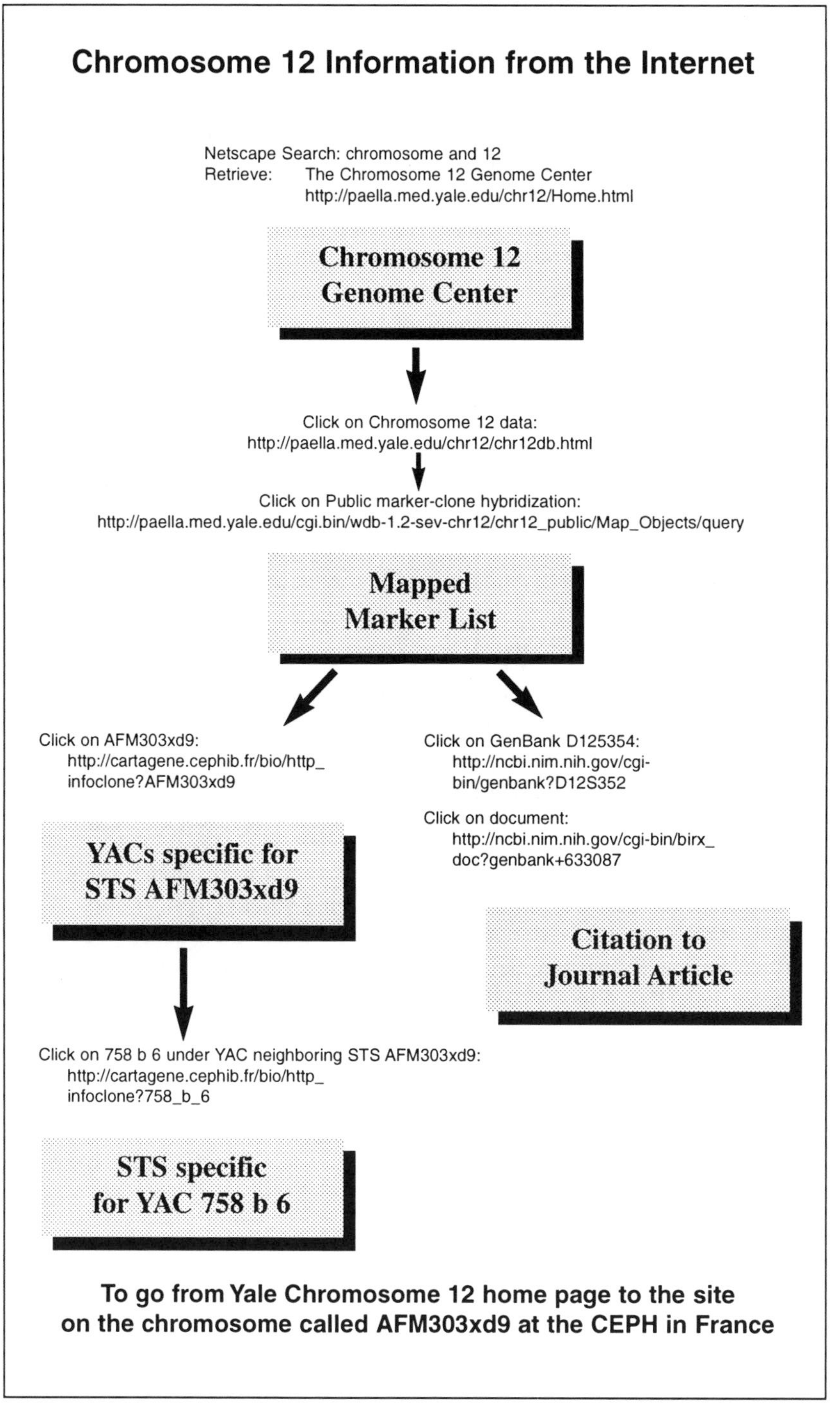

Chromosome 12 Information from the Internet

Netscape Search: chromosome and 12
Retrieve: The Chromosome 12 Genome Center
            http://paella.med.yale.edu/chr12/Home.html

Chromosome 12
Genome Center

Click on Chromosome 12 data:
http://paella.med.yale.edu/chr12/chr12db.html

Click on Public marker-clone hybridization:
http://paella.med.yale.edu/cgi.bin/wdb-1.2-sev-chr12/chr12_public/Map_Objects/query

Mapped
Marker List

Click on AFM303xd9:
    http://cartagene.cephib.fr/bio/http_
    infoclone?AFM303xd9

Click on GenBank D125354:
    http://ncbi.nim.nih.gov/cgi-
    bin/genbank?D12S352

Click on document:
    http://ncbi.nim.nih.gov/cgi-bin/birx_
    doc?genbank+633087

YACs specific for
STS AFM303xd9

Citation to
Journal Article

Click on 758 b 6 under YAC neighboring STS AFM303xd9:
    http://cartagene.cephib.fr/bio/http_
    infoclone?758_b_6

STS specific
for YAC 758 b 6

**To go from Yale Chromosome 12 home page to the site
on the chromosome called AFM303xd9 at the CEPH in France**

Figure 3-1

49

for each database. In his guide, Bishop (1994) included an appendix of resources that listed almost one hundred databases, software packages, and communication programs relevant to genetics information.

Tracing the changes in access to GenBank illustrates how policies have broadened as they have evolved. GenBank was initially accessed through a dial-up service. A password was needed, but anyone could request and receive one. Users were then given twenty minutes of free searching. Anyone who needed more time had to use a private service. Telnet was used for a while, then interactive searching was discontinued; now access to GenBank is obtained through an electronic mail service. CD-ROM versions are currently available, but the growth of Internet, the ever-increasing size of the databases, the time lag in getting information on CD-ROMs, the need for frequent updates, and the licensing issues make it unlikely that the CD-ROM format will continue. NCBI, in the first issue of its *NCBI News* newsletter, suggested that subscribers to the CD-ROM version of Entrez "might find alternative means of access . . . more convenient and economical." (BankIt . . . 1995, p. 3). The article listed several advantages of not using CD-ROM:

- alternative access is free, with no subscription charges,
- CD-ROM drives are not required,
- a small amount of local hardware is required,
- a broader sequence-related subset is available, and
- daily updates and three-dimensional structures will soon be available.

Similarly, BLAST software provides a client-server environment for searching GenBank directly. The software for searching BLAST can be "ftp-ed" by anyone. Authorin software permits authors to add their data directly to GenBank through the Internet and provides step-by-step instructions for inputting data (BankIt . . . 1995). Authorin will be updated with the point and click software called Sequin (Adamson and Casey 1994).

In another example of how access policies have changed, GDB now gives anyone with a Sybase license access to read-only copy of its files (Brandt 1993); but this also has changed with the most recent release of Sybase that moves the product to a client-server architecture and opens GDB to anyone. Annotator's WorkBench software is similar to Sybase for users of GSDB (Adamson and Casey 1994).

The progression in access to genetics information parallels the development of computers and telecommunications in recent years. WWW has given almost anyone with the technical "know-how" a way to access genetic databases. In practice, high powered computers equipped with modems and a certain level of user sophistication is needed to gain access. And in some cases passwords are needed. Installing a WWW presents a number of hurdles to overcome: software and the telecommunications need to work together; the Internet provider might not have a service orientation; the software needs to be understood; there must be a sufficient number of ports; or all interfaces need to be working together. While software language is getting more uniform and the Windows format is becoming fairly standard, this may, and is likely to, change. The most frequently asked questions at NLM's help desk no longer relate to search strategy, but to telecommunication problems.

Standardization still erects a hurdle for the HGP. In theory, all genetic databases should be linked. Sites on the maps of chromosomes should lead to the corresponding sequences. For example, a link is needed from GenBank and EMBL sequence information with map locations in GDB. In 1991 links did not exist; a plan was developed to make such links transparent to the users and permit them to move from maps to sequence information seamlessly (Pearson 1991). By 1993 SWISS-PROT had links to twelve different databases (Bairoch and Boeckmann, 1993). Software, such as PROPHET (Hollister 1988), offers interfaces to molecular databases and helps scientists with the manipulation and analysis of data through tables, statistics, and graphs. One goal of these links, in addition to the obvious one of making information easier to understand and access, is to minimize redundancy.

Changes in Communication Among Scientists

Traditional means of communication continue to be used; with the advent of electronic publishing, new methods are added to old. A paradigm shift is taking place. For a time, as predicted by Kuhn (1970), both old and new methods will be used simultaneously. Paper copies of journals are still the standard; online databases supplement the information in journals. Out of necessity, given the large amounts of data produced, the HGP has moved more rapidly than some other disciplines to an electronic environment. Data are too cumbersome and voluminous to be handled solely by the paper media.

The number of newsletters in this field has increased along with the number of listserves; these forms of communication exist side-by-side with some redundancy. *Human Genome News* began in 1989 and is published jointly by the NIH and DOE to facilitate communication among researchers. It is available at no charge to anyone interested in the HGP and can be obtained either directly over the Internet or by mail. An example of a more specialized newsletter is *Genome Digest* (HUGO . . . 1993) a quarterly newsletter of the European region, produced by HUGO with a focus on genome research in Europe. *NCBI News* and *DDBJ Newsletter* focus on the particular projects covered by these two programs. These and a number of other newsletters contain typical current

51

information: announcements of grants, fellowships, meetings, and projects; overviews of programs; and important news items.

Bleasby et al. (1992) claimed that one of the "best kept secrets" of the whole project is the availability of computer networks, stating that many networks were not used or known about to the extent that they should be. They listed twenty-nine news groups from BIOSCI, a nonprofit news group network, with step-by-step instructions on how to get "connected." A year earlier, however, Grefsheim, Franklin, and Cunningham (1991), in a study of how biotechnologists get and use information at a major academic center, found that computers had a significant impact on the way information was acquired. During the first full week of operation in December, 1993 the NCBI WWW site was accessed on average by thirty unique outside users each day. One and one-half years later, during the last full week of June, 1995, this average had increased to 1380 unique users per day. A total number of 251,169 unique host logon occurrences were posted for this two and one-half year period (National Center for Biotechnical Information 1995).

By 1993 DOE and NCHGR had identified four groups of users who would want or need information on the HGP: genome centers; laboratories at research institutions; small laboratories and individual investigators; and the public, including students, individuals, and those in small clinics (Snoddy and Robbins 1993). At the April, 1993 informatics meeting, attendees predicted that the "Internet . . . will become an essential highway for routine . . . data submission, retrieval, and analysis." (ibid, p. 2). It is indicative of how quickly progress is being made that this rather recent prognostication sounds so obvious in 1996.

Given the increasing amount of information on the Internet and its growing access, many are questioning the continued need for scientific journals. This topic was the subject of a June 1995 *Business Week* article, which said the $4 billion-a-year publishing industry could no longer afford to print scientific journals (Hamilton and Dawley 1995). A *BMJ* editorial in May 1995 called for scientific journals "to move beyond the vinyl records of journal articles to the CDs of research communication on the electronic information superhighway." (LaPorte et al. 1995, p. 1390)

The vast amount of data the HGP generates is changing a form of scientific communication that has been basically unaltered for 300 years. In the traditional model, new scientific discoveries are subjected to the peer review process before the information is distributed in print format, usually the journal literature. Data generated through the HGP are directly input into electronic databases. While the BMJ article acknowledges that peer review delays its intention of letting "all readers of the paper serve as reviewers" (ibid 1995, p. 1388) by attaching their comments electronically to the article, the method is not unlike the current system of publishing letters to the editor following publication. The

HGP is in the middle of a paradigm shift as it moves from a print to an electronic format.

Model of Invisible Colleges

Small groups of closely connected researchers ("invisible colleges") have traditionally exchanged raw data before those data or a summary of the data became available to the scientific community through the journal literature. We have seen that much of the raw data that comprises the HGP are now available to anyone who has access to the Internet. Information on the human genome has become an online extension of the laboratory bench.

The HGP has brought together a large number of people working together on one of the largest undertakings in the history of science. The modernized Garvey-Griffith Model developed by Hurd (Figure 3-2) can be applied to the HGP with a slight alteration. During editorial peer review, the reviewers have access simultaneously to both the raw data of the HGP and the manuscript that explains the data. After review and the accepted manuscript is published, the raw data remains in data sets that are available to anyone. Interested scientists and the public, in theory, now have access to information (raw data) that previously was available only to members of the invisible colleges, which, within the HGP, are not as closed as they had been. The researchers, students, and others interested in the HGP have a much easier time locating and communicating with scientists. The producers of GenBank redirect any question they receive to an appropriate researcher. They act as the conduit to get the requesters to the most likely source for the answer.

In the traditional model of scientific communication, membership in invisible colleges was limited to a small group of researchers. Anyone on the "outside" had little or no access to what was said, what was happening, or what decisions were being made. Conversely, the capabilities of electronic communication include, of course, the option of private e-mail communication and closed listserves.

Although the model of invisible colleges continues to exist today, within the context of an electronic environment information is less likely to stay within a small group of researchers. It is common to forward e-mail or listserve messages, thus removing the element of "closed access," if those in the invisible college communicate by e-mail. It is understood that much of genetic material is not confidential and can (and should) be read by anyone who might want it or need it. Databases are becoming more public then they once were. The government databases, which require that information be deposited in them if projects are federally funded, is another method of effectively removing old communication barriers. The nature and complexity of this work requires consider-

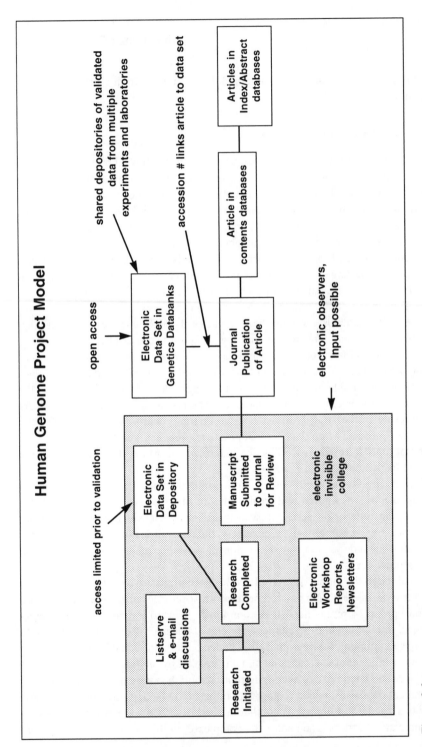

Figure 3-2

able interdisciplinary cooperation. Within the HGP, groups of invisible colleges are becoming much less invisible, altering the traditional model.

The challenges of this new form of communication are interesting: scientists must be connected to and must understand how to access these systems. The systems must be inexpensive to use and compatible, and standards are required. There must also be more interdisciplinary cooperation than ever before: among scientists, mathematicians, programmers, and computer experts (Pearson and Soll, 1991).

Unresolved Issues

Standards

Genetic databases are composed of four types of data: sequencing data, map locations, genetic diseases, and bibliographic references. Methods of integrating and presenting map information are problematic. Before full integration can occur among databases, these problems must be resolved. Because the HPG is carried out at institutions throughout the world, standards for language, indexing, mapping, keywords, and taxonomic classification to MeSH need to be developed; and submission procedures also need to be standardized. Systems must be designed to be used by "computer novices" and equipped with user friendly tools.

No universal standards exist for storing and representing map information. GDB staff members regard their attempt at storing and representing map information as experimental and, at the same time, realize that their experiment is setting the stage for more discussions and pilot projects (Pearson 1991). James Ostell, chief engineer at NCBI in 1993, argued for ASN.1 standards. He stated that he could name more than fifty molecular biology databases that do not communicate with each other (Aldhous 1993). Cox et al. (1994) called for a policy on universal measures that would outline a means to assess mapping progress. This policy could describe progress for any mapping project.

NCBI's goal is to produce an integrated system that would permit a researcher to retrieve relevant information from any existing system with one search. The quest for solutions to these problems is not new. PROPHET was first conceived in the 1960s as a program that would provide interactive tools for the study of chemicals and biological systems (PROPHET 1988).

Database differences

Inconsistencies and incompatibilities among databases is not a new issue. The nucleic acid sequencing databases actually differ slightly; however, this problem relates more to information added early in the project. Cuticchia et al. (1993) argued that there should be "one stop shopping" for the users of all the genetic databases. All databases should be seamlessly integrated into a single "virtual database," which would necessitate the universal adoption of a Standard Query Language (SQL). The authors admitted that while access is easier than it used to be, more work needs to be done. One hurdle to overcome is the inconsistency between databases due to different nomenclature for sequencing and mapping.

Borivoj Keil (1990) sounded an alarm by predicting that as long as sponsorship is not centralized, the number of databases will continue to increase. Effective cooperation and coordination among the various types of databases and the differences in the information they contain—including nucleic acid sequencing, protein sequencing, human genes, model organisms, literature, and mapping—is essential.

Print and electronic differences

Some authors request that electronic data not be released immediately. Information may then be altered during the review process, but those alterations may not be made to the electronic versions. Data not submitted electronically are keyed into a database manually—electronic scanning devices are not yet sophisticated enough to transcribe the data accurately. GenBank producers use two staff members as keyers who input all nonelectronic data GenBank receives. These two keyers are so accurate that their work is no longer checked; the input is shared immediately (Adamson and Casey 1994). But this activity leaves the potential for different print and electronic formats. It would not be clear to a user which version is correct. This problem is judged as minor by those working in the field.

Private Sector

The move to the private sector raises a number of issues: who owns the information, will the data continue to be free, or will restrictions to access become a problem? How would the private sector handle this information? A time may come when genetic information in a database could be traced to a specific individual. This opens the possibility of the private sector, especially insurance companies, using the information unethically (Murphy and Lappe 1994). The international aspect of the HGP dictates that the means of communicating need to remain open to all; and researchers will not be hurt because they do not have access to the full range of data and information that they need.

Long Term Benefits

Kuhn, in *The Structure of Scientific Revolutions* refers to certain scientific undertakings as "hack work" (Kuhn 1970, p. 30): mop-up work that can be relegated to engineers and technicians, but not "real" researchers. An early objection to the HGP was a concern that NSF would spend huge amounts of money on a project that was not basic research, and thereby divert funds from pure science researchers. These concerns have become less of an issue as long term benefits are starting to be realized. For example, the discovery of the specific genetic structure that might cause neurofibromatosis (the "elephant man's disease") could lead to a cure or prevention.

Conclusion

This chapter has examined a number of aspects of the HGP. It has put the HGP in an historical scientific context, reviewed informatics aspects, examined the model of editorial peer review, looked at access issues, and revised the model of invisible colleges.

When Margaret Dayhoff began a project to compile sequencing information as research tool, she probably could not foresee the long term outcome: worldwide databanks that have access to millions of pieces of information and that are updated nightly via satellite. Her first steps to organize the information were the beginnings of the informatics record keeping. The *Atlas of Protein Sequence and Structure* (Dayhoff et al. 1965) was published before computers played a significant role in the data's organization and dissemination. If all information on the DNA sequence of the human genome alone were printed, it would fill 1,000 volumes of 1,000 pages each (DOE, Office of Energy Research 1992), far more that Dayhoff's desire to print all sequences between a "single pair of book covers."

When Mendel introduced a new theory of inheritance, he had no colleagues to provide feedback on methodology, data interpretation, results, or conclusions. The three scientists who rediscovered his work forty years later did not have access to networks and to invisible colleges, or even an effective journal literature to help them place their research in the context of what had already been accomplished.

We are poised at the front edge of rethinking how scientific information is transmitted and who owns that information. Scientists no longer work in isolation or small groups, but work in collaboration among several disciplines. On an international scale, editorial peer review is no longer limited to two or three anonymous researchers who examine the manuscript before publication. Instead, the reviews are done by groups of chromosome experts who are appointed by an international committee and whose identities are known. These groups add a significant level of review to the traditional method of editorial oversight. Experts share information at a tremendous rate and work in an online scientific workbench opened to anyone to observe, comment on, or learn from. Modern communication techniques have forever altered the model of scientific knowledge sharing.

57

As of May 11, 1995, each of the nucleic acid sequencing databases stored information on 286 million base pairs from 352,000 sequences on humans and 8,000 other species. In that same month, a news item in *Nature* announced funding for designing pilot projects that would lead to the first full-length sequence of the human genome (Project to sequence . . . 1995). This announcement comes less than ten years after the Cold Spring Harbor meeting (see Table 3.2) where the merits, feasibility, and outcomes of such a project were debated.

Current estimates indicate that the HGP will cost less than originally predicted and be completed earlier than originally planned (Genome project . . . 1995). These accomplishments are due in large part to the increased speed and capacities of computers and of communications between them, the growth of networks, international cooperation, and international funding. All problems have not been resolved, nor has the full potential of such a project been realized, but the HGP will surely stand as one of the most significant accomplishments of the last decade of the twentieth century.

Abbreviations and Acronyms

CEPH (Centre d'Etude du Polymorphisme Humaine)
DDBJ (DNA Data Bank of Japan)
DOE (Department of Energy—U.S.A.)
EMBL (European Molecular Biology Laboratory)
GDB (Genome Data Base)
HGP (Human genome project)
HUGO (Human Genome Organization)
JIPIDS (Japan International Protein Information Database)
LANL (Los Alamos National Laboratory)
MIPS (Martinsried Institute for Protein Information Database)
NCBI (National Center for Biotechnology Information)
NCHGR (National Center for Human Genome Research)
NIGMS (National Institute of General Medical Sciences)
NIH (National Institutes of Health)
NLM (National Library of Medicine)
NSF (National Science Foundation)
OHER (Office of Health and Environmental Research)
OMIM (online *Mendelian Inheritance in Man*)
PIR (Protein Information Resource)
SQL (Structured Query Language)
URL (Uniform Resource Locator)
WWW (World Wide Web)

Definitions of Common Genetic Terms*

base pair (bp)—two nitrogenous bases (adenine and thymine or guanine and cytosine) held together by weak bonds. The human genome consists of 3 billion base pairs.

chromosome—the self-replicating genetic structures of cells. There are 46 chromosomes in 23 pairs in the human genome.

DNA (deoxyribonucleic acid)—the molecule that encodes genetic information. Four nitrogenous bases are arranged in base pairs along a sugar-phosphate backbone. The two strands of DNA are held together in the shape of a double helix by bonds between the base pairs.

gene—the fundamental unit of heredity. A section of the DNA located on a particular chromosome that encodes a specific function or physical attribute. There are more than 100,000 human genes.

genome—all the genetic material in the chromosome of a particular organism.

map—the relative positions of genetic loci (the position on a chromosome of a gene or other genetic marker) determined on the basis of how often the loci are inherited together.

nucleotide—a subunit of DNA consisting of a nitrogenous base (adenine, guanine, thymine, or cytosine), a phosphate molecule, and a sugar molecule.

polymerase chain reaction (PCR)—a means of "amplifying or making numerous copies of DNA in a test tube." (Roberts 1989)

recombinant DNA molecules—a combination of DNA molecules of different origin that are joined using recombinant DNA technologies.

sequencing—the order of nucleotide bases in a DNA molecule.

sequence tagged sites (STS)—a short (200 to 500 base pairs) DNA sequence that has a single occurrence in the human genome and whose location and base sequence are known.

* Unless otherwise noted, definitions from: *Human genome program. Primer on molecular genetics*. Washington DC: Department of Energy. Office of Energy Research. Office of Health Environmental Research, June, 1992.

References

Adamson, A., and D. Casey. 1994. Managing genome sequencing data. *Human Genome News* 6(3) September:1-6.

Aldhous, P. 1993. Managing the genome data deluge. *Science* 262(5133), October 22:502-3.

Avery, O. T., C. M. MacLeod, and M. McCarty. 1944. Studies on the chemical nature of the substance inducing transformation of pneumococcal types. Introduction of transformation by a deoxyribonucleic acid fraction isolated from pneumococcus type III. *Journal of Experimental Medicine* 79(1), January 1:137-57.

Bairoch, A., and B. Boeckmann, B. 1993. The SWISS-PROT protein sequence data bank, recent developments. *Nucleic Acids Research* 21(13):3093-6.

BankIt submissions mount. 1995. *NCBI News* (September). Available: http://www.ncbi.nlm.nih.gov/NCBI/news/sept95.html/

Barker, W. C., D. G. George, and L. T. Hunt. 1990. Protein sequence database. *Methods in Enzymology* 183:31-49.

Barker, W. C., D. G. George, H. W. Mewes, F. Pfeiffer, and A. Tsugita. 1993. The PIR-International databases. *Nucleic Acids Research* 21(13):3089-92.

Bishop, M. J. 1994. *Guide to Human Genome Computing*. London: Academic Press.

Bleasby, A., P. Griffiths, R. Harper, D. Hines, K. Hoover, D. Kristofferson, S. Marshall, N. O'Reilly, and M. Sundvall. 1992. Electronic communications and the new biology. *Nucleic Acids Research* 20(16), August 25:4127-8.

Brandt, K. A. 1993. The GDB Human Genome Data Base: a source of integrated genetic mapping and disease data. *Bulletin of the Medical Library Association*, 81(3) July:285-92.

Burks, C., M. J. Cinkosky, W. M. Fischer, P. Gilna, J. E. D. Hayden, G. M. Keen, M. Kelly, D. Kristofferson, and J. Lawrence. 1992. GenBank. *Nucleic Acids Research* 20 (Suppl):2065-9.

Burks, C., and L. J. Tomlinson. 1989. Submission of data to GenBank. *Proceedings of the National Academy of Science USA* 86 (January):408.

Cameron, G. N. 1988. The EMBL data library. *Nucleic Acids Research* 16(5):1865-7.

Cameron, G., P. Kahn, and L. Philipson. 1989. Journals and databanks. *Nature* 342 (December 21/28):8.

Casey, D. 1994. Genetic map goal met ahead of schedule. *Human Genome News* 6(4), 1 (November):14-15.

Chargaff, E. 1950). Chemical specificity of nucleic acids and mechanism of their enzymatic degradation. *Experientia* 6(6):201-9.

Chromosome editors. 1994. *Human Genome News* 6(2), July:8-9.

Cinkosky, M. J., J. W. Fickett, P. Gilna, and C. Burks. 1991. Electronic data publishing and GenBank. *Science* 252 (May 31):1273-7.

Cohen, S. N., A. C. Chang, H. W. Boyer, and R. B. Helling. 1973. Construction of biologically functional bacterial plasmids in vitro. *Proceedings of the National Academy of Science USA* 70(11), November:3240-4.

Collins, F., and D. Galas. 1993. October 1). A new five-year plan for the U.S. human genome project. *Science* 262 (October 1):43-6.

Cook-Deegan, R. M. 1989. The Alta summit, December 1984. *Genomics* 5:661-3.

Cox, D. R., E. D. Green, E. S. Lander, D. Cohen, and R. M. Myers. 1994. Assessing mapping progress in the human genome project. *Science*, 265(5181), September 30:2031-2.

Cuticchia, A. J., M. A. Chipperfield, C. J. Porter, W. Kearns, and P. L. Pearson. 1993. Managing all those bytes: the human genome project. *Science* 262 (October 1): 47-8.

Databases: limited access to BIONET. 1984. *Nature* 310(5980), August 30:717.

Dawid, I. B. 1989. Submissions of sequences. *Proceedings of the National Academy of Science USA*, 86:407.

Dayhoff, M. O., R. V. Eck, M. A. Chang, and M. R. Sochard. 1965. *Atlas of protein sequence and structure.* Silver Spring, MD: National Biomedical Research Foundation.

Dayhoff, M. O., R. V. Eck, M. A. Chang, M. A., and M. R. Sochard. 1967-68. *Atlas of protein sequence and structure.* Silver Spring, MD: National Biomedical Research Foundation.

Denson, D., M. Boguski, D. J. Lipman, and J. Ostell. 1990. The National Center for Biotechnology Information. *Genomics* 6(2), February:389-91.

Department of Energy. Office of Energy Research. Office of Health and Environmental Research. 1994. *Human genome. 1993 program report.*

Department of Energy. Office of Energy Research. Office of Health and Environmental Research. (June, 1992). *Human genome program. Primer on molecular genetics.* Washington DC: Department of Energy.

Department of Health and Human Services. Department of Energy. (1990). *Understanding our genetic inheritance. The U.S. human genome project: The first five years.* NIH publication No. 90-1580. National Center for Human Genome Research, National Institutes of Health.

EMBL Data Library, GenBank Staff. 1987. A new system for direct submission of data to the nucleotide sequence data banks. *Nucleic Acids Research* 15(18), September 25:ii-xii.

Ezzell, C. 1989. NIH to end support for electronic network. *Nature* 340, (July 13):87.

Five years of progress in the human genome project. 1995. *Human Genome News* 7(3 & 4), September-December:4-9.

Franklin, R. E., and R. G. Gosling, R. G. 1953. The structure of sodium thymonucleate fibres. I. The influence of water. *Acta Crystallography* 6:673-7.

Friers, W., R. Contreras, G. Haegeman, R. Rogiers, A. Van de Voorde, H. Van Heuverswyn, G. Volckaert, and M. Ysebaert. 1978. Complete nucleotide sequence of SV40 DNA. *Nature* 273(5658), May 11:113-20.

GenBank adding 1,000 human sequences per day from Merck project. 1995. *NCBI News* (March):1-2.

Genome project finishes fifth year ahead of schedule. 1995. *Human Genome News* 7(3 & 4), September-December:1.

George, D. G., W. C. Barker, and L. T. Hunt. 1986. The protein identification resource (PIR). *Nucleic Acids Research* 14(1):11-15.

Gilna, P., L. J. Tomlinson, and C. Burks. 1989. Submission of nucleotide sequence data to GenBank. *Journal of General Microbiology* 135:1779-86.

Gorner, P. 1994. Quicker DNA mapping expected. Argonne to use new technique to unlock human gene. *Chicago Tribune*, Sect. 1 (September 24):4.

Grefsheim, S., J. Franklin, and D. Cunningham. 1991. Biotechnology awareness study, part 1: where scientists get their information. *Bulletin of the Medical Library Association* 79(1), January:36-52.

Guyer, M. S., and F. S. Collins. 1993. The human genome project and the future of medicine. *American Journal of Diseases of Children* 147(11), November:1145-52.

Hamilton, J. O'C., and H. Dawley. 1995. Darwinism and the Internet. *Business Week* (June 26):44.

HNIH/CEPH Collaborative Mapping Group. 1992. A comprehensive genetic linkage map of the human genome. *Science* 258 (October 2):67-86.

Hollister, C. 1988. PROPHET—a national computing resource for life science research. *Nucleic Acids Research* 16(5):1873-5.

Holtzman, N. A. 1989. *Proceed with caution: predicting genetic risks in the recombinant DNA era.* Baltimore: The John Hopkins University Press. p. 5.

HUGO publishes *Digest.* 1993. *Human Genome News* 5(4), November:7.

Instructions to authors. 1983. *Journal of Biological Chemistry* 258(1), January 10:3.

Instructions to authors. 1984. *Journal of Biological Chemistry* 259(1), January 10:3.

Kahn, P., and D. Hazledine, D. 1988. NAR's new requirement for data submission to the EMBL data library: information for authors. *Nucleic Acids Research* 16(10):i-vii.

Keil, B. 1990. Cooperation between databases and scientific community. *Methods in Enzymology* 183:50-60.

Koshland, D. E., Jr. 1987. Sequencing the human genome. *Science* 236(4801), May 1:505.

Kristofferson, D. 1987. The BIONET electronic network. *Nature* 325(6106), February 2:555-6.

Kuhn, T. S. 1970. *The Structure of Scientific Revolutions.* Enlarged edition. Chicago: The University of Chicago Press.

LaPorte, R. E., E. Marler, S. Akazawa, F. Sauer, C. Gamboa, C. Shenton, C. Glosser, A. Villasenor, and M. Maclure. 1995. The death of the biomedical journal. *BMJ* 310(6991), May 27:1387-90.

Maddox, J. 1989. Making good databanks better. *Nature* 341(6240), September 20:277.

Martin, N. J., T. Primich, and R. A. Riley. 1994. Accessing genetics databases. *Database* (February):51-7.

Maxam, A. M., and W. Gilbert. 1977. A new method for sequencing DNA. *Proceedings of the National Academy of Science USA* 74(2):560-4.

McGourty, C. 1989. Who's hiding primary data? *Nature* 341(6248), September 14:94.

Murphy, T. F., and M. A. Lappe. 1994. *Justice and the Human Genome Project.* Berkeley: University of California Press. p. 94.

National Center for Biotechnical Information. 1995. July 20. (On-Line). Available: http//:www.ncbi.nlm.nih.gov.

National Center for Human Genome Research. 1995. Mission statement and organization.

Office of Technology Assessment. Congress of the United States. 1988. *Mapping our genes: genome projects: how big? how fast?* Baltimore: Johns Hopkins University Press. pp. 11-17.

Olson, M., L. Hood, C. Cantor, and D. Botstein. 1989. A common language for physical mapping of the human genome. *Science* 245 (September):1434-5.

Pauling, L., and R. B. Corey, R. B. 1950. Two hydrogen-bonded spiral configurations of the polypeptide chain. *Journal of the American Chemical Society* 72(11), November:5349.

Pearson, M. L., and D. Soll. 1991. The human genome project: a paradigm for information management in the life sciences. *FASEB Journal* 5(1), January:35-9.

Pearson, P. L. 1991. The genome data base (GDB)—a human gene mapping repository. *Nucleic Acids Research* 19(Suppl):2237-9.

Pearson, P. L., B. Maidak, M. Chipperfield, and R. Robins. 1991. The human genome initiative—do databases reflect current progress? *Science* 254(5029), October 11:214-5.

Project to sequence human genome moves on to the starting blocks. 1995. *Nature* 375(6527), May 11:93-4.

PROPHET, a national computing resource for life science research. 1988. *Nucleic Acids Research* 14(1):21-4.

Reddy, V. B., B. Thimmappaya, R. Dhar, K. N. Subramanian, B. S. Zain, J. Pan, P. K. Ghosh, M. L. Celma, and S. M. Weissman. 1978. The genome of Simian Virus 40. *Science* 200(4341), May 5:494-502.

Rice, C. M., R. Fuchs, D. G. Higgins, P. J. Stroehr, and G. N. Cameron. 1993. The EMBL data library. *Nucleic Acids Research* 21(13):2967-71.

Roberts, L. 1988. NIH and DOE draft genome pact. *Science* 241(4873), September 23:1596.

———. 1989. New game plan for genome mapping. *Science* 245 (September 29):1438-40.

Ruddle, F. H., and K. K. Kidd. 1989. The human gene mapping workshop in transition. *Cytogenetics and Cell Genetics* 52:1-2.

Sanger, F. 1956. The structure of insulin. In: Greene, D. E., ed. *Currents in biomedical research*. New York: Interscience Publishers. pp. 434-59.

Sanger, F., A. R. Coulson, T. Friedmann, G. M. Air, B. G. Barrell, N. L. Brown, J. C. Fiddes, C. A. Hutchison III, P. M. Slocombe, and M. Smith. 1978. The nucleotide sequence of bacteriophage 0X174. *Journal of Biology and Medicine* 125(3), November 5:225-46.

Sinsheimer, R. L. 1989. The Santa Cruz workshop—May 1985. *Genomics* 5:954-6.

Smith, T. F. 1990. The history of the genetic sequence database. *Genomics* 6(4), April:701-7.

Snoddy, J., and R. Robbins, R. 1993. Bioinformatics "highway" needed. *Human Genome News* 5(3), September 9:1-4.

Walker, R. T., M. P. Deutscher, J. E. Donelson, S. C. R. Elgin, R. J. Roberts, S. Tilghman, and W. R. McClure. 1987. Deposition of nucleotide sequence data in the data banks. *Nucleic Acids Research* 15(18), September 25:i.

Watson, J. D. 1990. The human genome project: past, present, and future. *Science* 248(4951), April 6:44-9.

Watson, J. D., and F. H. C. Crick. 1953. Molecular structure of nucleic acids. A structure for deoxyribose nucleic acid. *Nature* 171(4356), April 23:737-8.

Woodsmall R. M., and D. A. Benson. 1993. Information resources at the National Center for Biotechnology Information. *Bulletin of the Medical Library Association* 81(3), July:82-4.

Yager, T. D., D. A. Nickerson, and L. E. Hood. 1991. The Human Genome Project: creating an infrastructure for biology and medicine. *Trends in Biochemical Science* 16(12), December:456-61.

Chapter Four

High Energy Physics

Julie M. Hurd

It is clear that the world of physics is on the verge of a revolution, a revolution that is driven by technology, but whose true nature will be determined by the response of the world scientific community. The revolution will change what and how physicists read, how they become aware of what they read, and even what "read" means.

—Report of the APS Task Force (1991)

High energy physics, also known as particle physics, employs the very largest of experimental equipment—particle accelerators—to study the smallest objects in the universe—subatomic particles. This discipline is an important branch of contemporary physics because it attempts to answer very fundamental questions about the nature of the universe. High energy physicists study matter on a sub-atomic scale and probe the nuclei of atoms to describe the elementary particles that serve as basic building blocks for the universe and the forces that bind these particles together.

Early in the twentieth century physicists believed that atoms were the small-est discrete particles of matter. Studies of radioactivity, cosmic rays, and elec-tromagnetic radiation were soon to disprove that belief. British physicist Ernest Rutherford is credited with the discovery of the atomic nucleus, which he and co-workers "observed" by bombarding gold foil with high-energy radioactive particles. Rutherford concluded that the atomic nucleus is the location of prac-tically all the mass of an atom even though it represents a very small portion of the space that that atom occupies. Rutherford's research contributed to a new model of the atom (known as the Bohr model, after Danish physicist Niels Bohr), which consists of a dense nucleus of protons surrounded by orbiting electrons. Rutherford also hypothesized the existence of another elementary nuclear particle, the neutron, although it was not until 1932 that James Chadwick would prove experimentally the existence of neutrons.

While Rutherford had employed radioactive particles to bombard nuclei and induce decay into constituent particles, physicists who followed him turned to cosmic rays emanating from outer space as sources of higher energy beams. Cosmic rays entering the earth's upper atmosphere collide with atoms present there and break them into constituent particles. The resulting cascade of smaller particles produces secondary particles and these are analyzed when they reach the earth's surface. Particle physicists use cloud chambers and bubble chambers to study the tracks the particles produce. From these studies came the discovery of positrons and other antimatter particles. (Antimatter particles are short-lived counterparts of identical mass but opposite charge to matter; for example, a positron is a positively charged counterpart to the electron. When matter and antimatter particles come together, they annihilate each other converting their mass to pure energy.)

During the 1920s, physicists first began to construct equipment to produce even higher energy beams for the study of subatomic particles. They built particle accelerators, devices that subject charged particles such as protons and electrons to electric and magnetic fields in order to raise their energies. Some accelerators speed particles along very long straight tunnels; these are the "linear accelerators" such as the machine at Stanford University, one of the largest in existence. Other accelerators—cyclotrons and synchrotrons—employ a circular track; the synchrotron at Fermilab in Batavia, Illinois comprises a ring about four miles in circumference and is one of the most powerful machines of its type.

The highly-energized particle beams emerging from accelerators are used to bombard target atoms, and the resulting collisions produce the elementary particles that physicists identify and analyze. Today physicists believe that all matter is made up of twelve elementary particles, six known as leptons, of which one is the electron, and six others called quarks; combinations of these basic particles form larger particles such as protons and neutrons.

The first particle accelerators were small and built with private funding, but as physicists aspired to produce higher and higher energy beams, they sought ever larger and more powerful machines. Government funding became increasingly important in the construction of large accelerators such as those at Brookhaven National Laboratory on Long Island, the University of California—Berkeley, and Stanford University. Typically, Atomic Energy Commission grants supported the construction of an accelerator and its supporting facilities and paid the salaries of many of the scientists who staffed the laboratories (*The Ultimate Structure of Matter* 1990). High energy physics certainly fits Price's description of Big Science (1963), with its costly experimental facilities and large teams of scientists that publish numerous multiple-authored research papers.

As high energy physics developed and scientists learned more about the fundamental particles that make up the universe, the knowledge gained found applications in other fields. The particle physics experiments of the 1920s led to theories of quantum mechanics that were crucial in the development of transistors and lasers in the 1940s and 1950s. More recently, advanced medical technologies have borrowed from basic high energy physics research and identified new uses for both the experimental devices and the techniques used to analyze data. Applications derived from physics provide powerful therapeutic and diagnostic tools that are becoming commonplace in modern health care facilities.

Radiation therapy for the treatment of cancer employs equipment that is based on the concepts used to construct the first particle accelerators. Beams of charged particles are used in high-precision surgery and serve as scalpels in sensitive and delicate operations. For example, proton beams have become the preferred treatment for tumor growths inside the eye; the precision with which a proton beam can be controlled makes it less damaging to surrounding tissues than a surgical scalpel. Other important medical applications include Computerized Axial Tomography (CAT scanners) and Magnetic Resonance Imaging (MRI scanners) that permit noninvasive observations of soft tissues. MRI and CAT scanners provide powerful imaging capabilities and allow medical scientists to "view" soft tissues and organs within the human body. These now-essential diagnostic devices draw on the equipment and mathematical techniques first developed by particle physicists.

Additional applications of high energy physics research can be found in other fields, including food science, the electronics industry, and the transportation industry. Products now common in daily use such as microwave ovens, air traffic control radar, and photocopiers, owe their existence to research that originally set out to answer basic questions about the nature of matter. There is no doubt that high energy physics has had a significant impact on most citizens of the world's developed countries.

Communication in Physics: The Role of Preprints

As in all scientific disciplines, communication in physics includes both formal and informal elements. The traditional communication system in physics that has evolved over the last century corresponds to Garvey and Griffith's basic model. The behavior of physicists as information seekers and users has been studied extensively over the last several decades by numerous sociologists and information scientists. For example, the communication behavior of physicists, particularly in their use of journals, was described by Donald King and Nancy Roderer

(1982). They identified the circulation base of a journal and the speed of publication as the most important attributes by which physicists judged journals, whether to read or to serve as channels of publication for their own research.

Physicists are most interested in current literature and much less likely to use older materials than scientists in many other disciplines. King and Roderer reported "separates" to be an increasingly important form of distribution; separates include reprints, preprints, and photocopies of articles. The use of preprints is a noteworthy feature that plays a significant role in shaping the future system of communication in high energy physics where the exchange of preprints has long been an established part of the discipline's culture. Preprints are, as their name implies, precursors of articles that may eventually appear in a research journal and, in high energy physics, exchange of preprints among members of invisible colleges has long been a key element in informal communication.

Garvey and co-workers (1970, 1972) studied the information exchange processes associated with the production of scientific journal articles and found extensive use of prepublication reports, including preprints. Physical scientists were more active in the distribution of preprints than social scientists or engineers. Over half the authors in their study distributed at least one preprint prior to publication of their articles. More authors shared preprints before submitting the manuscript to a journal, but some waited until the manuscript had been submitted for editorial consideration or even until receiving notification of acceptance. Six was the median number of preprints distributed; however, one-sixth of the authors reported sending out twenty-five or more preprints.

Distribution lists for preprint mailings were found to be variable. Three-fourths of the authors surveyed by Garvey sent preprints to colleagues working in the same area. Smaller numbers of authors mailed to contract or grant distribution lists, institutional mailing lists or formally organized preprint exchange groups. Preprint distribution demonstrates an aspect of invisible college activity that serves several purposes for those involved. Early dissemination of findings is only one benefit; opportunities to receive feedback appeared to be important for many authors who reported making both stylistic and substantive changes in manuscripts based on comments received.

Warren Hagstrom (1970) examined use of preprints in forty-six scientific specialties in four scientific disciplines, one of which was physics. He was able to correlate high rates of preprint distribution with specialties at very active research fronts. These specialities produce literature that becomes obsolete rapidly, a characteristic of physics literature in general. Among the specialties within physics that he studied, "nuclear physics" and "elementary particles" had the highest rates of preprint distribution. Hagstrom suggested that preprint distribution may be part of the distinctive culture of physicists.

Ruth Kramer (1985) examined the use of preprints by scientists and synthesized findings from a number of important studies of the medium, including those of Garvey, Hagstrom, and King and Roderer. She, too, identified physics as a discipline in which the preprint appeared to be particularly important to early informal communication of research.

In her case studies, Kramer described libraries that serve physicists and astronomers. Here, preprint collections are organized and controlled through indexing and filing systems that provide subject and author access to collections that number in the thousands. She cited an early effort at the Stanford Linear Accelerator (SLAC) to announce and speed preprint distribution for the particle physics community that was supported initially by an Atomic Energy Commission grant. The SLAC announcement list was later automated with National Science Foundation support as part of the development of Stanford Public Information Retrieval System (SPIRES). High energy physicists clearly value the SLAC service, which eventually became self-supporting and is used by high energy physics laboratory libraries worldwide.

All of the studies cited above provide evidence for the importance of preprints to scientific communication among physicists, particularly in rapidly developing specialties such as particle physics. The studies describe an environment that should surely be particularly receptive to applications of emerging information technologies that hold promise for improving access to preprints. It is not surprising, therefore, that users have readily adopted an electronic preprint distribution system developed for the high energy physics community by Los Alamos physicist Paul Ginsparg. The electronic preprint archive maintained by Ginsparg now serves specializations in physics, computational linguistics, economics, and other disciplines; and it serves potentially as a model for other areas of study in which preprints are an important communication tool.

Electronic Preprints

Paul Ginsparg (1994) described the history of the "e-print archives" that he developed and continues to operate from a Los Alamos Laboratory computer. The first database, hep-th (for High Energy Physics—Theory), was planned for use by a community of about 200 physicists and was first available in 1991. Ginsparg acknowledged that the existence of a "preprint culture" in high energy physics that predated electronic communication networks certainly contributed to a positive response to his efforts.

Ginsparg identified concurrent developments in computer software and hardware that were essential to the construction of electronic preprint databases. First, he cited the acceptance during the mid-1980s by the physics community of the software application known as TeX; its scientific word processor standardized the manuscript creation process. The second essential development was the great increase in computer connectivity that gave rise to the Internet. Finally, high-powered workstations with high-capacity storage media became available.

ASCII-based TeX supports file transfer across various hardware platforms and provides for generation of figures. TeX also allows scientists to produce a print manuscript with all the formatting features of a journal publication. Distance collaboration is thus freed from the constraints imposed by the need to mail drafts of manuscripts; instead manuscripts in process could be available in real time to collaborating authors on computers connected to a network.

Ginsparg saw potential in the emerging technologies for development of an electronic preprint archive and distribution network. He wrote supporting software to automate the processes that would allow users to submit and replace papers, search and obtain preprints, and receive online assistance in using the system. He incorporated a current awareness function that allows users to subscribe and receive a daily listing of titles and abstracts of new papers added to the database. He intended that minimal computer literacy would be required and designed his system around an e-mail interface.

Ginsparg assembled an initial distribution list for hep-th from two e-mail distribution lists totaling 160 addresses In six months the list had grown to over 3600 subscribers (Ginsparg, 1994). By 1994 the database received approximately 200 new manuscripts per month, responded to more than 700 e-mail requests daily, and, on peak days, transmitted more than 1000 copies of e-prints. This distribution service requires about 70 megabytes of storage per year (at a 1994 cost of $70) and is claimed to place a negligible burden on Los Alamos Laboratory's network resources. There is no charge to users of the distribution service. The software appears to run with little intervention and can be left unattended for extended periods.

Physicists' acceptance of e-print archives has already begun to impact their libraries. The Special Libraries Association's Physics/Astronomy/Mathematics Division sponsors an active electronic discussion list (SLA-PAM@ listserver.lib.muohio.edu) that has explored the issues related to electronic versus printed preprints. During a series of discussions on the list during 1995, a number of major physics libraries reported that they had discontinued their paper preprint collection and organization efforts, relying instead on the e-print archives; other libraries continued to maintain paper files but had noted decreased usage and were monitoring developments closely (STS-L, 1995).

A significant difficulty cited by some librarians relates to the desire of users to print copies of papers of interest from the archive. Not all physicists presently have equipment on their desktops that can handle graphics and mathematical notation; compressed graphics files that required UNIX machines for decompressing appeared to be a particular problem. Those who had access to suitable equipment, however, whether through their library or elsewhere in the institution, were pleased with the quality of prints they could obtain. It seems likely that, as more physicists upgrade their workstations, these early difficulties will diminish.

The initial success of this pioneering e-print archive has served as a model for other databases. The software running on the Los Alamos computer in Paul Ginsparg's office now supports sixteen different distribution lists in physics and six in mathematics, as well as several in other specialized areas. While the system was not planned as an alternative to scientific periodicals, it appears to serve as an electronic journal for many users. It offers value-added features such as a speedier distribution system that is open to a wider community of readers than paper-based preprint distribution and brings the latest research to readers who are distant from well-stocked libraries or who lack personal subscriptions to journals. This system supports "electronic invisible colleges" of specialists (provided, of course, they possess suitable computer equipment and a connection to the Internet) who can share research findings and discuss common interests wherever they are located.

At this time the e-prints are still "preprints" and virtually all are submitted to refereed journals for traditional publication. The e-print archive software permits authors to insert a citation to a published article at any time, although not all necessarily do this. Additionally, the SLAC Library tracks published papers in the electronic index of preprints that they have maintained for over two decades, and Ginsparg (1996) reports that he "has downloaded over 10,000 of the published records directly from their database."

Hrvoje Galic (1996) of the SLAC Library provided the information in Table 4.1 on the fate of e-prints in high energy physics. Galic also observed that many of the remaining e-prints that lack citations to published articles probably end up in books or proceedings, but the SLAC Library does not try to track this information.

Statistics maintained at Los Alamos document the usage of the e-print archives and show steady growth from their inception. During the first week of 1996 there were 134,926 connections from 9,923 hosts (E-print Archives 1996).

Rikki Lewis (1995) has identified trends in the use of reprints and discusses several variables that may affect traffic in reprints. The ready availability of large research library collections with low-cost photocopiers has led to diminished requests for reprints from the developed countries, but scientists without access to large research libraries and/or inexpensive copy facilities continue to make heavy use of reprints. Lewis reports that some authors give priority to requests from scientists in developing countries. Availability of e-prints offers potential for faster and less costly access for those scientists—again, provided they have access to a suitable workstation and an Internet connection.

Some physicists, who acknowledge the value-added filtering provided by journal editors and referees, have expressed concerns related to quality control

Table 4.1

Journal Publication of High Energy Physics E-prints

Archive	Total E-prints	Total Linked to Published Articles	% Published
hep-ph (est. 3/92) (phenomenology)	8,093	4,265	52.7
hep-th (est. 9/91) (theory)	8,640	4,778	55.3
hep-lat (est. 2/92) (lattice)	1.595	818	51.3
gr-qc (est. 7/92) (general relativity and quantum cosmology)	1,963	911	46.4

of submissions to e-print archives. Paul Ginsparg has argued that the widespread distribution of the archive fosters a self-policing check on quality that may serve to limit premature submissions and incorrect results; the potential for global embarrassment is far greater in a networked environment than in a paper-based distribution to a few selected colleagues! Ginsparg has also suggested ways that a refereeing function might be implemented by employing selected expert reviewers prior to distribution or by volunteer readers after posting. He suggests the possibility of partitioning the database into reviewed and unreviewed sectors as a guide to readers.

Transformed Models for Communication in Physics

The e-print archive that Paul Ginsparg developed as an experiment is likely only the first step in a process that promises to lead to a transformed system of scientific communication in physics. Figure 4.1 offers a model of such a system that draws on the hep-th experience and incorporates a reviewing function as Ginsparg has proposed.

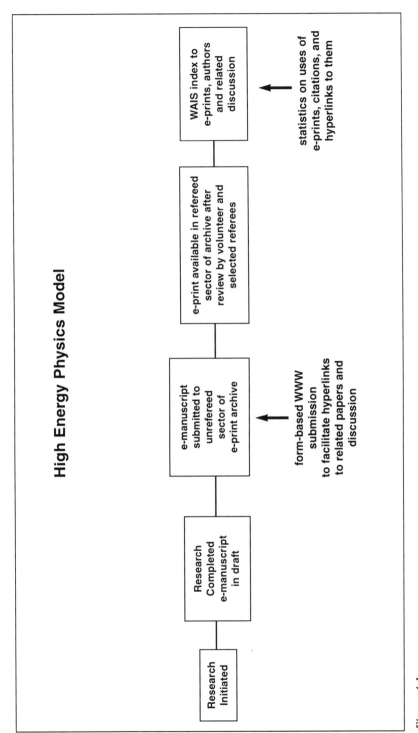

High Energy Physics Model

Research Initiated

Research Completed e-manuscript in draft

e-manuscript submitted to unrefereed sector of e-print archive

form-based WWW submission to facilitate hyperlinks to related papers and discussion

e-print available in refereed sector of archive after review by volunteer and selected referees

WAIS index to e-prints, authors and related discussion

statistics on uses of e-prints, citations, and hyperlinks to them

Figure 4-1

This model is a "journal-less" paradigm with the e-print as the basic unit of distribution. Just as specialized journals collected articles for their readers, e-print archives in specialized fields would play the same role of partitioning a large body of literature. This model blurs the distinction between informal communication as represented by preprints and formal communication as in published journal articles. It also shifts roles of authors and publishers: authors become "publishers" by the act of transmitting an e-print to the database. The model offers potential for a more broadly-based type of peer review as any reader of an e-print may comment on a submission. Whether this electronic-based distribution scheme will eventually displace traditional refereed journals is yet to be determined, but the high use that has been measured for the high energy physics prototype provides evidence of the value physicists place on the service.

It is worth noting that the American Physical Society (APS) several years ago envisioned a model of an electronic information system based on articles as basic units of distribution. The APS formed a task force in 1988 that was charged with reviewing current and projected technological developments related to storage, retrieval, and delivery of physics information. The group was also directed to develop a strategy, a time scale, and a plan to utilize information technologies to supply physics information more effectively. After deliberations lasting more than a year, the task force released a lengthy report containing its "Vision for the Year 2020" for a transformed physics information system. The report described the hardware and software environments, the nature of the "new" literature, how it would be produced, disseminated, and used, and how older "pre-electronic" information would be managed (Report of the APS Task Force 1991).

Central to the APS report is a "Physics Information System," an electronic physics library containing all published physics information centrally stored, searchable online, and continually updated. Documents in the database would be linked to related documents and to relevant comments and discussion, creating "an entire web of referencing" forward and backward in time. The timing of this report predates the World Wide Web, digital library initiatives, and the hep-th e-print archives, but these more recent developments seem to connect the APS vision statement to an emerging reality that closely reflects the intent of the APS authors.

Some of the unresolved problems cited in the report, notably those pertaining to technological limitations, have already been solved. Others that relate to economic issues and the role of publishers, for example, remain as determinants in a paradigm shift that is underway but not completed.

Since the APS report appeared, several publishers of physics journals have responded to opportunities for alternative distribution schemes and initiated their own experiments. The World Wide Web promises to be a particularly attractive means of presentation of electronic journals because it supports graphics and multimedia exchange of information.

Major association and commercial publishers of physics journals are rapidly establishing Web pages, and use them in a variety of ways. Association publishers, such as the American Institute of Physics (AIP), communicate with their memberships through their Web pages. AIP has made several of its free newsletters available on the Web and is planning additional electronic publications. The Institute of Physics (IOP), London, has promised to make all of the journals published by its not-for-profit subsidary, Institute of Physics Publishing, available on the World Wide Web by Spring 1996.

Subscription journals can be distributed over the Web because the servers that run Web software can recognize Internet addresses of paid subscribers. In this way a library that subscribes to Institute of Physics journals can arrange for affiliated users of the library to access the IOP server and read materials on their own or library workstations. IOP plans to experiment with distribution of preprints of articles accepted for publication in its journals.

Commercial publishers, such as Elsevier, are also planning electronic publications and are experimenting with use of networks to distribute tables of contents, preprints, and full text. It seems apparent that the next few years will see more experiments with electronic distribution. The successes and failures of these efforts will shape the communication system of the future

Numerous questions need answering that relate to acceptance of technology, changing reward systems, the role of association and for-profit publishers, copyrights, archiving, and more. Some of these issues are addressed in more detail in the final chapter. Evolving technology may render some concerns moot, and new questions not yet articulated may emerge. Whatever the future holds for communication in physics, however, it is apparent that the shift from printed paper to electronic media is well underway.

References

E-print Archives. 1996. (January 25). Available: http://xxx.lanl.gov/cgi-bin/show-stats)

Galic, H. 1996, January 18. electronic mail communication.

Garvey, W. D., N. Lin, and C. E. Nelson. 1970. Some Comparisons of Communication Activities in the Physical and Social Sciences. In Nelson, C. E., and D. K. Pollack, eds. *Communication among Scientists and Engineers*: 61-84. Lexington, MA: Heath Lexington.

Garvey, W D., N. Lin, and K. Tomita. 1972. Research Studies in Patterns of Scientific Communication: III. Information Exchange Associated with the Production of Journal Articles. *Information Storage and Retrieval* 8: 202-221.

Ginsparg, P. 1994. First Steps Towards Electronic Research Communication. *Computers in Physics* 8 (No. 4): 390-396.

———. 1996. Electronic mail communication. January 17.

Hagstrom, W. O. 1970. Factors Related to the Use of Different Modes of Publishing Research in Four Scientific Fields. In Nelson, C. E. and D. K. Pollack, eds. *Communication among Scientists and Engineers*: 85-124. Lexington, MA: Heath Lexington.

King, D. W. and N. K. Roderer. 1982. Communication in Physics—the Use of Journals. *Physics Today* (October): 43-47.

Kramer, R. 1985. *The Role of the Preprint in Communication among Scientists*. Educational Resources Information Center: ED261685.

Lewis, R. 1995. Reprints Still Enabling Authors to "Spread the News" about Their Work. *The Scientist* (December 11): 16-17.

Price, D. J. de Solla. 1963. *Little Science, Big Science*. New York: Columbia University Press.

Report of the APS Task Force on Electronic Information Systems. 1991. *Bulletin of the American Physical Society* 36: 1119-1137.

STS-L summary of SLA-PAM discussion. 1995. (April 22). Available: sts-1@utkvm1.utk.edu

The Ultimate Structure of Matter: the U.S. High Energy Physics Program from the 1950s through the 1980s. 1990. Washington, DC: U.S. Department of Energy, Office of Energy Research.

Chapter Five

Astronomy, Astrophysics, and Space Physics

Susan Y. Crawford

Some 1.5 gigabytes of data are beamed to the earth each day from the Hubble Space Telescope. As of mid-1994, the Hubble archive contained more than 2,000,000 files in 400,000 data sets that occupied 1.1 terabytes of storage space on optical disks. The lifespan of Hubble, estimated at approximately 15 years, is but one of the great space observatories scheduled to be launched during the last decade of this century. Such is the magnitude of data management in the space sciences today.

—Andernach, Hanisch, and Murtagh (1994)

Until 1957 when the former Soviet Union launched its Sputnik, astronomy was a relatively small field in which both amateurs and professionals pursued their fascination. During the Cold War, which fueled fierce international competition, the discipline mushroomed into the "space sciences," with strategic concerns and practical applications. In 1958 the National Aeronautics and Space Administration (NASA) was established, replacing the National Advisory Council for Aeronautics. The new agency, with greatly expanded funtions, would conduct research for solving the problems of flight and for exploring space with manned and unmanned vehicles (U.S. Government Manual 1995/96). Putting spacecraft into the air, landing a man on the moon, and tracking the orbits of artificial satellites were among the new objectives.

Today, the space sciences entail both pure science and application. Aeronautics, astronautics, and meteorology are based on such pragmatic goals as national defense, sources for new raw materials, biological survival in space, and monitoring Earth's environmental conditions. Astronomy and astrophysics attempt to explore the larger picture of the cosmos through observations that are collected, analyzed, and integrated into the body of scientific knowledge. They address basic processes such as the origin and structure of the universe, the formation of galaxies and planetary systems, the collapse of massive stars to form black holes, and the beginning of life forms in the universe. As shown in Figure 5.1, on scientific programs of NASA, it is difficult to separate pure research from its application and core areas from related areas (ibid).

Astronomy is largely concerned with collecting data and is interdisciplinary in approach. In very early times, the field was heavily algebraic—concerned with planetary movement relative to the position of stars and constellations. Later, it was found that the laws of physics on Earth—concerned with gravity, matter, and energy—also held for the universe, and astronomy and physics became allied sciences (Novak 1995). As the objective is to discover laws about the universe, astrophysics embraces many areas, among them chemistry, computer science, geology, material sciences, engineering, and biology.

With the passage of Cold War politics that had impelled much of our government's interest in space exploration, followed by the budget cutting initiatives of the current Congress, astronomy is again experiencing problems. In early 1996, such costly and esoteric research as exploring distant galaxies attracts little Congressional interest. Even the earth's near collision with asteroids in the last three years (look what happened to the dinosaurs!) and the dramatic meeting between Jupiter and Comet Shoemaker-Levy 9 have stirred only transient interest among the general public. In 1995, NASA's $14 billion budget was slated to be cut by $2 billion over the next five years, which included an expected reduction in the agency's work force from 23,000 to 17,000 (Lawler 1995; Lawler, 1996). NASA reported, in its 1994 strategic plan, that the "changing environment has led us not only to restructure programs in development, but also to fundamentally alter our approach to future missions." (National Aeronautics and Space Administration 1994)

This chapter focuses on information handling in astronomy, astrophysics, and space physics and does not separate theoretical from applied research. For serious professionals, the field is data intensive, fast moving, technology dependent, and costly "Big Science."

Scientists were interviewed at the American Astronomical Society (Washington, D.C.), the NASA Goddard Space Flight Center (Greenbelt, Maryland), the Harvard-Smithsonian Center for Astrophysics (Cambridge, Massachusetts), the University of Michigan Upper Atmospheric Research Collaboratory (Ann Arbor, Michigan), Northwestern University Department of Physics and Astronomy (Evanston, Illinois), the University of Chicago Department of Astronomy and Astrophysics (Chicago, Illinois), and Yerkes Observatory (Williams Bay, Wisconsin).

78

Scientific Programs of the National Aeronautics and Space Administration, 1995

SPACE SYSTEMS DEVELOPMENT

To develop large-scale space systems from conception to operational readiness, e.g., Space Station Freedom, a permanently manned space station

SPACE SCIENCE AND APPLICATIONS

To understand the origin, evolution, and structure of the universe, the solar systems, and the integrated functioning of the earth, e.g., space physics and solar system exploration

SPACE FLIGHT

Space flight operations involving humans, cargo habitable space facilities, e.g., development and testing of operations of the space shuttle program

AERONAUTICS AND SPACE TECHNOLOGY

To develop advanced technology in pursuit of national objectives in aeronautics, space, and transatmospherics, e.g., High Performance Computing and Communications Program

EXPLORATION

To consolidate the nation's space exploration efforts to conduct human and robotic missions to the moon and Mars

SPACE COMMUNICATIONS

To provide global communications system links tracking sites, control centers, and data processing facilities that provide real-time data processing for mission control, orbit and attitude determination, and routine processing of telemetry data for space missions

Figure 5-1

Data Collection and Management in the Space Sciences

A new project, the Sloan Digital Sky Survey, is being planned using a 2.5 meter telescope at Apache Point Observatory in New Mexico. The telescope will continuously slew across a region of the sky north of the Milky Way— about a quarter of the celestial sphere. Its objective is to glean further clues into the structure of the universe. To that end, it will plot the positions of some 50 million galaxies and about 70 million stars. The census of the skies will be made not with photographic plates, but by using silicon detector chips that generate electric charges when they are exposed to light. The greatest challenge, according to some astronomers, is not in developing the telescope, but in the technology for digitally recording the enormous number of images and for automating their analysis, recognition, and classification (Hayes 1994).

The Sloan survey underscores two aspects of Big Science today: the magnitude of the projects and the formidable problems of information handling. These and similar developments in other areas of science emphasize the impact of technology and Big Science on the scientific communications system.

There are four major steps in the data collection, organization, management, and transfer system of the space sciences:

1. data are collected by ground-based or air-borne instruments
2. data are tabulated in observation logs, online databases, and catalogs
3. Universal Resource Locators and coordinated databases are developed to manage and to locate the data
4. data are transmitted via networks to computer workstations.

Data Collection

Until some 20 years ago, astronomical observations were made almost entirely from ground-based, optical observatories usually built on high mountain peaks. Among others, the gigantic telescopes at Mt. Wilson (100-inch) and Mt. Palomar (200-inch) in California and at Mauna Kea, Hawaii continue to further our understanding of the universe. But observations from the earth's surface are impeded by atmospheric conditions that act as a veil and tend to distort incoming signals. Additionally, scientists are concerned not just with images that emit or reflect light, but also with signals that emanate from invisible or unknown sources. Powerful observatories have therefore been placed in orbit above the earth's atmosphere.

In the United States, four "great observatories" have been planned over the early 21st century, each directed to detecting a different part of the electromagnetic spectrum: gamma rays, x-rays, visible and ultraviolet light, and infrared rays. Two have so far been launched: the Gamma Ray Observatory (GRO) and

80

the Hubble Space Telescope (National Aeronautics and Space Administration, Astrophysics Division, n.d.).

In 1990, a space shuttle carried the Hubble Space Telescope (HST) into orbit some 370 miles above the earth, from where its builders hope it will observe the universe for 15 years or longer. Every three years or so, astronauts will be sent to Hubble to fix or to replace parts. Aiming at a galaxy, planet, star or other object, HST receives signals that are then directed to scientific instruments on board. The signals are beamed to a computer network on the ground where the analysis takes place. Computers display the signals as photographs, spectra, and other types of data.

It is important to note that astronomical research today is not confined to massive programs such as the large NASA or European Space Agency (ESA) initiatives. Individual investigators, based at universities and observatories, pursue their own research agendas, often requesting time and data from these large facilities.

Tabulation of Data–Astronomical Catalogs and Online Databases

Data in the space sciences are collected by many centers, ranging from international to private programs. The development of an integrated gateway to the main data centers has not yet been achieved. The Astrophysics Data System (ADS), a NASA initiative, attempted to provide a common user interface to a wide variety of data sets, but this proved not feasible due to lack of consensus among the main users (Kurtz 1996). Some data collected by active research centers are kept at those centers, while other data, especially from the large national programs, are deposited into archives. In contrast, the Human Genome Program has a mechanism for mandatory deposit of data. When a manuscript is submitted for publication, the major genome-related journals have agreed that the supporting data must be electronically deposited into internationally shared databases (see Chapter 3). Unfortunately, such is not the case in astronomy and astrophysics today.

Andernach, Hanisch, and Murtagh (1994) distinguish the following databases:

- *Astronomical catalogs* that represent static, final compilations for a given set of objects. Two centers—the Astronomical Data Center (ADC) of the NASA Space Science Data Center (NSSDC) and the Centre de Données Astronomiques de Strasbourg (CDS), France have systematically coordinated many catalogs into "results" databases. CDS archives some 900 astronomical catalogs and has put its most recent catalog on the World Wide Web. The scope of ADC, broadly defined, is "all data of interest to the NASA community," from federally-funded initiatives to Project VEGA (Russian data on Venus).

81

- *Observation logs and/or archives of raw or calibrated data* from astronomical observatories, many of which can be accessed online. Examples include data from the Hubble Space Telescope, the Extreme Ultraviolet Observer, and the Infrared Space Observatory.
- *Online Databases via Internet* that manage and coordinate data from a number of sources. Chief among these is SIMBAD (Set of Identifications, Measurements, and Bibliography for Astronomical Data) compiled by CDS, a log of over 1.03 million astronomical objects outside the solar system. EINLINE, at the Harvard/Smithsonian Center for Astrophysics, provides access to over 50 databases related to the Einstein X-ray Satellite archive. NED, the NASA/IPAC database, gives positions, names, and basic data for some 250,000 extragalactic objects and includes 380,000 bibliographic references.

For a comprehensive, international list of astronomical catalogs, online databases, observatory data, and other network resources for astronomers, Andernach, Hanisch, and Murtagh (1994) is an excellent source. Nonetheless, the authors cautioned that it may be the last published listing of its kind, as relevant information is changing so fast and increasing amounts of information are now offered on network tools such as the World Wide Web. Updates will be available on the Web.

Universal Resource Locators and Coordinated Databases

When an investigator completes a project, the findings are still reported in a scientific paper. The Astrophysics Data System (ADS) enables scientists to search this literature by providing access to a database of abstracts to journal articles and proceedings. The abstracts are linked to sources where the full-length articles may be retrieved through the World Wide Web. ADS now has over 800,000 abstracts online and is expanding to include full-length articles from the major journals.

Beginning October 1995, the Space Telescope Science Institute and the Harvard-Smithsonian Center for Astrophysics are coordinating a new service called "ASTROBROWSE" in cooperation with a number of astronomical organizations, among them NSSDC, University of Massachusetts, Goddard Space Flight Center, and the Centre de Données Astronomiques. ASTROBROWSE is intended as a system that allows astronomers and educators to find astronomical information and data by querying various data centers simultaneously through a common interface. It is a successor of the ADS program, whose objective was to search data from various data centers and to retrieve them. As of November 27, 1995 there were no published reports readily available on ASTROBROWSE, but information may be found by contacting its home page at http://niit1.harvard.edu/AstroBrowse (Eichhorn 1995).

The management of databases is an acknowledged problem in the space sciences. Because the many data collections and information services have not been integrated so that they can be accessed through a single gateway, a universal resource locator (URL) has been developed to guide users to sources. Astronomy Web Resources (ASTROWEB), supported by a consortium of seven institutions, is a compilation that provides addresses for making links to the following:

- Guides, directories
- Catalogs, online databases
- Object-oriented resources
- Telescope and observatory archives and home pages
- Software archives
- Journals, newsletters, preprint sources
- Library and information resources.

ASTROWEB does not provide data, but indicates where the resources may be located. Under "Observing Resources," for example, one may find, as of 1995:

- Telescopes (185 records)
- Astronomical survey projects (19 records)
- Telescope observing schedules (24 records)
- Meteorological information (14 records).

By clicking on the desired category, the user will find sources and addresses displayed.

Data Transmission

Data sets may be accessed or transferred directly to a scientist's computer workstation. Flam (1993) reported plaintively:

> Astronomers using the Hubble Telescope miss out on some of the romance of traditional astronomy: the lonely nights on a mountaintop, adjusting and angling a big telescope to zero in on a suddenly interesting nebula or galaxy. Instead, they observe without leaving their offices. First, they fill out paperwork more complex than income tax forms and send it off; then they settle in for weeks or months; finally they receive in the mail a tape bearing their data.

As indicated, there are often difficult problems and protocols to be overcome. The data coming to the ground daily from remote sources need to be put online and so are not available immediately. Leisavitz and Cheung (1995) at the Goddard Space Flight Center reported that it is not always easy to get the data collected by scientists on NASA projects. When a project is completed and the results published, the data are often "thrown into boxes or drawers" where they are forgotten. One of their functions is getting such data back into the NSSDC archives.

The analysis and management of data in a distributed network environment also pose formidable problems. To provide scientists with the ability to search for algorithms, applications programs, or utilities from the many software packages in use, directories of software archives have been developed by ASTROWEB and by Hanisch and co-workers (Andernach, Hanisch, and Murtagh 1994). The Science Information Systems Interoperability Conference, held at the University of Maryland in November 1995, is one of a series of international meetings to discuss problems in data management and to demonstrate systems (Science Information Systems 1995). The challenges of data analysis and integration have been extensively discussed by Albrecht and Egret (1991), Heck and Murtagh (1993) and Crabtree, Hanisch, and Barnes (1994).

Figure 5.2 shows the many sources of data that are available to space scientists from their workstations. These include observations online, astronomical archives, preprints, publications, directories, news groups, home pages, announcements, and small special interest groups or invisible colleges.

ASTROWEB is a universal resource locator that lists resources in astronomy and provides addresses for locating them. ADS (Astrophysics Data System) searches abstracts from major journals and guides the user to sources for full length papers. ASTROBROWSE is a new system under development to enable querying of various data centers simultaneously through a common interface. SIMBAD is a coordinated database that is object oriented; it is the prime source for data and for references to publications on astronomical objects.

Research Organization and Information Management in the Space Sciences

Rather than a single model of research organization in the space sciences, at least three (or their combination) can be identified. The objectives and the nature of research differ among the models, which in turn affect the way that information is managed and used. The three models consist of the following:

- *Massive federal initiatives*—some with international collaboration, funded by NASA, the Department of Defense (DOD), the National Science Foundation (NSF), and other agencies such as the U.S. Geological Survey and the National Institutes of Health (NIH).
- *Individual initiatives*—principal investigators at universities, research institutes, and observatories who work with a small staff of assistants and graduate students. These scientists have initiated research proposals or responded to requests for proposals, usually from NASA and NSF.

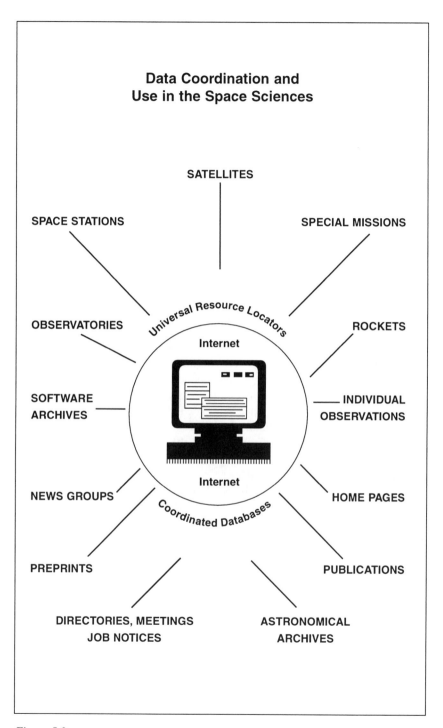

**Data Coordination and
Use in the Space Sciences**

SATELLITES

SPACE STATIONS

SPECIAL MISSIONS

OBSERVATORIES

Universal Resource Locators

ROCKETS

Internet

SOFTWARE
ARCHIVES

INDIVIDUAL
OBSERVATIONS

Internet

NEWS GROUPS

Coordinated Databases

HOME PAGES

PREPRINTS

PUBLICATIONS

DIRECTORIES, MEETINGS
JOB NOTICES

ASTRONOMICAL
ARCHIVES

Figure 5-2

85

This is the model on which the Garvey/Griffith studies (Reports of the American Psychological Association, 1963, 1965, 1969) were based.

- *Consortia or group initiatives*—often multidisciplinary and from different institutions, that collaborate on programs or projects. Research is also funded in large part by NASA and NSF, with some assistance from private foundations.

Generally, massive *federal initiatives* such as the NASA programs tend to be mission oriented, multidisciplinary, international in scope, and large-scale; in short, they are characteristic of Big Science enterprises as Julie Hurd describes them in Chapter 2. Social policy of the time and the political climate heavily influence which scientific programs will be undertaken. Work is largely done through contracts with industry and with research and development organizations. A single program may take more than a decade to develop and involve many contractors and specialities, e.g., computer technology, missile systems, and vehicle structures. Heavy bureaucratic oversight is required to administer such programs.

The data are collected for specific objectives and are often directed to a particular set of users. The proposed, permanently-manned space station "Freedom" will provide technology and support to the Department of Defense and will assist the private sector in identifying potential applications and commercial uses for NASA-developed technology. Publication in scientific journals for both the scientists and the funding agencies is an important objective in federally sponsored, targeted research. As in the case of the Hubble Space Telescope, data sets become public domain and individual investigators may reserve time for use of the instrument.

Individual initiatives, usually based in universities or institutes, play an important role in education as well as research. Principal investigators carve out areas of specialization and seek funding to support their work and their students or assistants. Research tends toward exploring, testing hypotheses, and collecting data on space phenomena, such as star formation and gas flows around black holes and neutron stars.

Consortia or group initiatives, such as the Sloan Digital Sky Survey and the Center for Astrophysical Research in Antarctica, are programs that entail multidisciplinary groups from several institutions. Some research and development are done in response to requests for proposals from federal agencies while others may be supported by several sources, including the private sector. For participants in these massive, group programs publication is also an important goal both for priority in discovery and for survival in the academic world.

The scientific communications system investigated by Garvey and Griffith was based on individual and small group initiatives, primarily in university settings. The study was not concerned with large government-funded, often international collaborations, which derive from national objectives. How does the university-based scientific communications process of today differ from the Garvey/Griffith model of the 1960s? Figure 5.3 shows the current sequence of steps from initiation of a project through the dissemination of results in academic space sciences.

The basic steps—project initiated, research completed, manuscript prepared, report published, and report indexed—remain essentially unchanged, as in Hurd's modernized Garvey/Griffith model. What has changed relates to modern technology—the use of electronic methods and communication networks to manage and to transfer the information in each of the steps.

When the research project begins, publications and data archives are searched electronically. New work may be tagged according to individual interest profiles so that scientists can keep up with developments at the frontiers. Under development are attempts to make an integrated gateway for retrieving information in a few steps, regardless of location. Hyperlinks will enable users to connect with papers in related fields or with resources in databases. Through the Internet, scientists can discuss their work with colleagues in other research centers throughout the world.

Scientists collect data through observation time on one of the large telescopes or other institution-based facility. Alternatively, data beamed from space-borne or ground-based observatories, such as the Hubble telescope or the International Ultraviolet Explorer, may be tapped online for analysis, comparison, or supplemental data. Data may be downloaded into local computers and analyzed by shared software. As noted by Eichhorn (1995), "the Web makes life much easier—you can see data from observations at your desktop, you can give instructions to the [telescope] operator, and you can transfer images." Scientists situated remotely from each other may form "collaboratories" to work together on problems (see last sections of this chapter).

Preliminary findings are still reported to granting agencies and to colleagues at conferences. New observations continue to be reported to the Central Bureau for Astronomical Telegrams at the Harvard-Smithsonian Center for Astrophysics in Cambridge, Massachusetts and other coordinating agencies. But the Bureau now issues electronic telegrams to its subscribers to announce discoveries; e.g., the sighting of a new nova, comet, or other object.

Manuscripts are submitted electronically for publication, abstracts and preprints are announced via newsgroups on the Internet, and full preprints with graphics may be received by e-mail. As reported by *New Astronomy*, papers are increasingly issued in electronic format, with the printed version published solely for archival purposes (Taubes 1996). The electronic version may include motion pictures or video simulations to show, for example, the impact on Jupiter of Comet Shoemaker-Levy 9 or the birth of new stars in a galaxy.

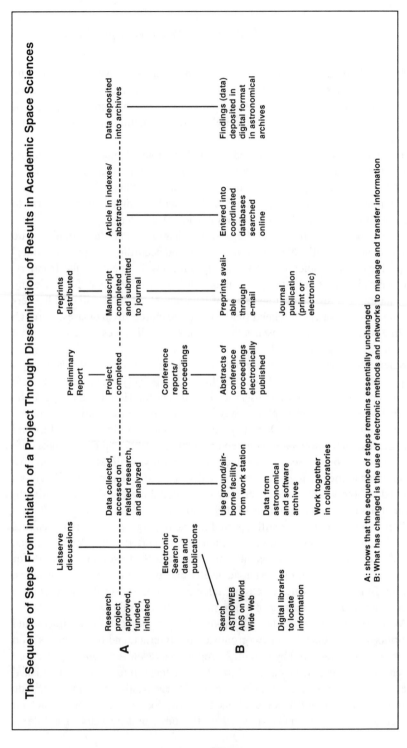

The Sequence of Steps From Initiation of a Project Through Dissemination of Results in Academic Space Sciences

A: shows that the sequence of steps remains essentially unchanged
B: What has changed is the use of electronic methods and networks to manage and transfer information

Figure 5-3

Electronic dissemination greatly increases the power to communicate and to investigate.

Retrieval is usually made through a search of abstracts in the Astrophysics Data System, which directs the user to a source for data or for a full article. During the last third of 1995, some 16,000 users accessed ADS, which is now used routinely. The data collected may be kept in active data centers or deposited into astronomical archives in the form of intermediate tabular data; both are accessible to scientists through the World Wide Web. As we shall see in the next section, however, rather than merely speeding up data collection and reducing costs, information technology will bring about increasing changes in both the organizations for information management and the methods of scientific practice.

Collaboratories

As previously noted, the sequence of steps in producing and communicating information in astronomy and astrophysics has not changed markedly over the past thirty years. Meanwhile, the scientific enterprise continued to grow and the volume of scientific information doubled every twelve years, straining the traditional publications system. Likewise, the cost of journals more than doubled between academic years 1985-86 through 1992-93 (Stix 1994). Getting research findings peer reviewed and published in the most prestigious journals presented unacceptable time lags. To meet these challenges, electronic publication spontaneously blossomed to include over 100 peer-reviewed, scientific journals by the end of 1995 (Taubes 1996).

Given profound changes in technology and knowledge of how scientists communicate and work, new methods for information handling have appeared in a number of specialties such as brain research, the Human Genome Project (see Chapter Three), and particle physics (see Chapter Four). The Human Brain Project plans to use computer networks and software to integrate databases in the neuroscience community, from the molecular and cellular level through brain mapping and cognitive psychology (Gibbons 1992). The Center for Astrophysical Research in Antarctica reports a project in which twenty collaborators communicate electronically on the development of a telescope, discussing such details as the number of pits that can be tolerated on a lens (Novak 1995).

Serving as a catalyst, the National Science Foundation began two initiatives: digital libraries and the organization of "collaboratories." The National Research Council defines a collaboratory as a "center without walls, in which the nation's researchers can perform their research without regard to geographical location—interacting with colleagues, accessing instrumentation, sharing data and computational resources (and) accessing information in digital libraries." (National Research Council 1993). Funding for collaboratories was awarded to six organizations, among them, the University of Illinois for scientists studying cell differentiation in the worm, *C. Elegans*, and the University of Michigan for scientists studying the earth's upper atmosphere.

The Upper Atmospheric Research Collaboratory

Scientists in the Upper Atmospheric Research Collaboratory (UARC) at the University of Michigan are concerned with turbulence on the surface of the sun as it affects the earth. As the outer atmosphere of the sun continuously blows away, a "solar wind" of extremely high temperature travels through interplanetary space where it distorts the earth's magnetic field and ionizes its upper regions. These effects on the earth's magnetosphere, ionosphere, and upper atmosphere are measured and studied by radar and a variety of instruments (University of Michigan UARC 1995a).

The UARC project started in 1992 and will conclude in 1997. It involves real-time instruments and investigators—the community of scientists who use ground-based instruments at the Sondrestrom Upper Atmosphere Facility in Kangerlussuaq, Greenland. Initially, participating scientists were situated at five sites: the Danish Meteorological Institute in Copenhagen, the University of Michigan, the University of Maryland, SRI International in Menlo Park, California, and Lockheed Palo Alto Research Laboratory. Later, sites at Cornell University, Florida Institute of Technology, University of Alaska, and University of New Hampshire were added.

Objectives of UARC

The UARC user community serves as a test bed for the development of electronic environments that will support team collaboration. UARC will provide scientists at various locations with access to each other, to remote facilities, and to information bases (digital libraries).

The project intentions and goals consist of the following:

- Combine computing and communication technology to create this experimental environment
- Investigate the longitudinal effect of the collaboratory on scientific research and on the scientists
- Apply this model to other academic projects and to other types of organizations such as international corporations and enterprises
- Assess whether computing environments such as UARC will become important components of the National Information Infrastructure (NII) initiative.

Test Bed Architecture

The first 18 months involved development of the infrastructure with a team composed of computer scientists, domain (space) scientists, and psychologists.

As of September 1995, the project was in its fifth major redesigning based on experience and feedback.

Initially, the project was in "wire service" where a stream of data are generated from instruments, collected, stored, and displayed at all user sites. This is evolving to a "point and talk" phase in which researchers can interact from different sites through shared windows, shared annotations to data and graphics, electronic drawing boards.

Figure 5.4 gives an overview of the UARC client architecture that provides the following services (University of Michigan UARC 1995b):

- Data viewing services: to view desired data from instruments in appropriate formats
- Core group services: to define new groups, joining a group, communication among members of a group
- Window sharing services: to share interfaces to data, e.g., export/import windows
- Replay services: to annotate data, play back archival data, replay experiments, collaborate via session display capture and replay
- Collaboration services: to discuss among users, to audio-conference, and to edit
- Information services: to look up and access various data sources, including digital libraries, the World Wide Web, and people sources.

Implications for Scientific Communication

In a collaboratory, the physical location evaporates. According to Clauer (1996), associate director for the UARC user community, the "participants are brought together as if they were sitting around a table in Greenland." Scientists can communicate with each other around the clock, observe and receive data from the Greenland facility, discuss findings, and add annotations.

The architecture is intended to support a large number of collaborative groups distributed across wide-area networks. It is anticipated that groups of participants will be small—around twelve persons—who will come together to solve problems and then disband. Collaboratories function much like electronic invisible colleges.

Digital Libraries

In 1994, in tandem with the collaboratory test bed, the University of Michigan has embarked on another NSF-funded project—the digital library. The underlying technology of digital libraries is based upon powerful communication networks and the transmission of digitized data across them. Today, information encoded in digital formats can be represented by electronic, optical, and magnetic phenomena and be transmitted and processed near the speed

The UARC Client Architecture
Consists of a Core Collection of Application

DATA VIEWING SERVICES

Enable user to view desired data in appropriate formats from instruments, e.g. radar, all sky imager

GROUP SERVICES

Include *window-sharing service* that allows users to share their interfaces to data (viewer windows) so windows can be exported and imported among users

Include means for defining new groups, communication among members, joining groups, exiting from groups.

REPLAY SERVICES

Allow users to annotate data for later reference, to replay scientific experiments, and to collaborate, via session capture and replay

COLLABORATION SERVICES

General purpose tools for a collaboration environment, e.g., multi-user *chat* windows that enable discussion among users, group editing, and audio conferencing

INFORMATION SERVICES

Provide lookup and access to various data sources, including people. A name service identifies optimal locations for services

Services Designed to Accommodate Needs of the Scientists

Figure 5-4

of light and stored at atomic-scale densities (University of Michigan Digital Library Project 1995a).

The data in digital libraries are viewed as objects in an information space connected by relational links (Epstein and Shakes 1995; Fox et al. 1995). The objects may be electronic representations of documents, data types (such as genes and chromosomes), or audiovisual media (such as images and sound). The concept may even be applied to the collection of things without physical analogs, such as algorithms, real-time data feeds, computational states, and relationships among versions of a physical object (Fox et al. 1995). The links and objects create an information space that is much like points within a web. Users navigate this space to seek interlinked information that may be found in many locations.

In 1994, a \$24.4-million Digital Library Initiative was started by NSF, NASA, and ARPA (Advanced Projects Research Network). Funding was awarded to six organizations, among them the University of Michigan. The objective of the University of Michigan Digital Library Project (UMDL) is to plan "a coordinated program of experimental research and development to gain insight into the creation, operation, and use of advanced digital libraries" (University of Michigan Digital Library Project 1995b). A digital library test bed is being developed with focus on the earth and space sciences.

UMDL incorporates a large number of distributed information sources and communicates over existing networks such as the Internet. It is a modular system that uses a variety of "agents" to coordinate different tasks such as query processing, information integration, and information management. When a user requests information, an interview agent determines the parameters of the search and an interface agent helps to formulate queries and to select appropriate information. This integration of distributed, disparate systems, with many data formats, currently poses many difficult problems that are now being addressed.

UMDL aims to provide interactive searching and browsing across many remote databases simultaneously and automatically, without the user having to select particular databases. The support structure and the location of the information becomes transparent to the user. The digital library will be accessible through a variety of computer workstations and networks. Users will also be able to add their documents to the databases, a publishing feature that will be implemented with controls.

The research team for the project encompasses many issues and areas of expertise, as follows:

- Electrical engineering and computer science: distributed intelligence systems—object-oriented databases, distributed artificial intelligence, collaboration systems

- Library and information studies: user needs, information organization, searching, and retrieval
- Publishing and intellectual property issues
- Design and construction of systems, human/computer interface design
- Socioeconomic and psychological factors.

The university's department of Atmospheric, Oceanic, and Space Sciences is an active participant in this experiment on integrating information digitally and processing queries to the system. Additionally, partnerships have been forged with many organizations. Elsevier and University Microfilms are among the publishers that are supplying journals in digital format.

Discussion

Vast amounts of data are collected, processed, archived, and used in astronomy and astrophysics. Over the past 30 years, computers and communication networks have profoundly changed how information in the space sciences is managed and transferred. Major journals and proceedings are now published in both electronic and hard print formats. But the basic steps of the communication sequence are still in place—online data gathering, intermediate tabular data, and peer reviewed publication. The linking of research data with publications and the interface of databases to provide an integrated gateway remain problems.

Two new research initiatives in the space sciences will speed the trend toward electronic information exchange. The digital library and the collaboratory are two components of a new scientific communication system being forged at the University of Michigan. The test beds, focused on upper atmospheric research, are worth watching—they suggest implications for both the space sciences and for the larger scientific enterprise.

The collaboratory is an electronic extension of the "invisible college" (Price 1986) that will bring scientists together to work on problems and to share the facilities of Big Science without traveling to distant sites. The digital library will enable scientists to transcend organizational walls and to access and explore data from many sources through hypertext links. Nevertheless, the implications for information management, for the user, for publishing, and for education are now being investigated, and we are barely starting to understand what these electronic tools can do (Boyce 1995). At the Upper Atmospheric Research Collaboratory, scientists conclude that "the emergence of collaboratories represents an important convergence of computing with scientific practice" but caution that it is yet too early to know their effect on the process of doing science. (Finholt and Olson 1995).

The final chapter discusses the implications of the new technology as observed in the three fields—genome sequencing, high energy physics, and the space sciences.

References

Albrecht, M. A., and D. Egret, eds. 1991. *Databases and Online Data in Astronomy.* Dordrecht, the Netherlands: Kluwer.

Andernach, H., R. J. Hanisch, and F. Murtagh. 1994. Network resources for astronomers. *Publications of the Astronomical Society of the Pacific* 106 (November):1190-1208.

Boyce, P., personal communication. American Astronomical Society, November 2, 1995.

Clauer, A., interview with author. Upper Atmospheric Research Collaboratory, University of Michigan, Ann Arbor, February 1996.

Crabtree, D. R., R. J. Hanisch, and J. Barnes, eds. 1994. *Astronomical Data Analysis Software and Systems III*, ASP Conf. Ser. No. 61. San Francisco, CA: ASP.

Eichhorn, G., interview with author. Harvard-Smithsonian Center for Astrophysics, Cambridge, MA., November 24, 1995.

Epstein, H., and D. Shakes, eds. 1995. C. Elegans: modern biological analysis of an organism in cell biology. *Methods in Cell Biology*: Vol. 48.

Finholt, T. A., and G. M. Olson. 1995. From laboratories to collaboratories: a new organizational form for scientific collaboration. Ann Arbor, MI: University of Michigan (preprint).

Flam, F. 1993, June 18. Space telescope institute: inside the Black Box. *Science* 260:1716-17.

Fox, E. A., R. M. Akscyn, R. K. Furuta, and J. J. Leggett, eds. 1995. Digital libraries. *Communications of the ACM* (April). Special issue on Stanford University.

Gibbons, A. 1992. Databasing the brain. *Science.* 258 (December 28): 1872-1873.

Hayes, B. 1994. Scanning the heavens. *American Scientist* 82 (November-December): 512-516.

Heck, A., and F. Murtagh, eds. 1993. *Intelligent information retrieval: the case of astronomy and related space sciences.* Dordrecht, The Netherlands: Kluwer.

Hevly, B. 1992. Reflections on big science and big history. In Galison, P., and B. Hevly, eds. *Big science and the growth of large scale research.* Palo Alto, CA: Stanford University.

Kurtz, M., interview with author. Harvard-Smithsonian Center for Astrophysics, February 7, 1996.

Lawler, A. 1995. Will NASA's research reforms fly? *Science, 271* (November 17). 1108-1110.

Lawler, A. 1996. Goldin puts NASA on new trajectory. *Science*, 272 (May 3): 800-802

Leisavitz, D. and C. Cheung, interview with author. Goddard Space Station, Maryland, August 29, 1995.

National Aeronautics and Space Administration. 1994. *Space science for the 21st century; strategic plan for 1995-2000.* Washington, DC: NASA, Office of Space Science.

——— n.d. *The great observatories for space astrophysics.* Washington, DC: NASA, Astrophysics Division.

National Research Council. 1993. Committee on a National Collaboratory Establishing the User-Developer Partnership. *National collaboratories: applying information technology for scientific research.* Washington, DC: National Academy Press.

Novak, G., interview with author. Dearborn Observatory, Northwestern University, June 1995.

Price, D. 1986. *Little science, big science . . . and beyond.* New York: Columbia University Press. 56-81.

Reports of the American Psychological Association's Project on Scientific Information Exchange in Psychology. 1963, v.1; 1965, v.2; 1969, v.3. Washington, DC: American Psychological Association.

Science Information Systems Interoperability Conference. 1995. University of Maryland Conference Center, College Park, Maryland, November 6-9.

Stix, G. 1994. The Speed of write; trends in scientific communication. *Scientific American*, 276 (December):106-111.

Taubes, G. 1996. Science journals go wired. *Science*, 271 (February 9):764-765.

University of Michigan Digital Library Project. 1995a. *Digital library project research proposal. 2.1, Background and Introduction.* On-line. (Available: http://http2.sils. umich.edu/UMDL/indexes/document-index.html)

University of Michigan Digital Library Project. 1995b. *Digital library project research proposal. 1.0, project summary.* On-line. (Available: http://http2.sils.umich. edu/UMDL/indexes/document-index.html)

University of Michigan Upper Atmospheric Research Collaboratory. 1995a. *The earth's magnetosphere.* On-line. (Available: http://WWW.sils.umich.edu/UARC/HomePage. html)

———. 1995b. *Design considerations in building a distributed collaboratory.* Ann Arbor, MI: University of Michigan School of Library and Information Studies.

U.S. Government Manual. 1995/96. Washington, DC: Government Printing Office, 1995: 607-612.

Chapter Six

The Changing Scientific and Technical Communications System

Julie M. Hurd, Ann C.Weller, and Susan Y. Crawford

Alan Revson is an associate professor of astrophysics at a large Midwest University. At eight o'clock one memorable morning, he sits at his desk scanning numbers on his computer in the 8-feet by 10-feet office that he shares back-to-back with a graduate assistant. On this morning, Revson is studying the clouds of gas around black holes that exist millions of light years distant from our solar system. When he pauses to check his e-mail, however, he finds one message that he has eagerly awaited: the Space Telescope Science Institute in Baltimore has scheduled two time slots for him to use the Hubble Space Telescope. Revson feels a surge of excitement, as only about ten percent of proposals to use the Hubble are approved (Flam 1993).

Before the night of Revson's scheduled observation, his instructions are transferred by cable to the NASA-Goddard Space Flight Center in Greenbelt, Maryland, then bounced to the telescope. Within 48 hours, scientists are processing the data that have been transmitted back to a ground-based station. A tape with the calibrated data is forwarded to Revson.

For months, Revson will analyze this data, determine its relationship to other observations, and question whether his findings fit into existing theory. He will then present the results at the annual meeting of the American Astronomical Society, and later publish them in a journal that is issued both in paper and electronic formats. An abstract of his paper will be entered into the Astrophysics Data System, which scientists can search on the Internet. Finally, the data from Hubble are deposited into data archives that may be accessed worldwide. Most of the observations, data analysis, consultations with colleagues, and paper preparation are done in Revson's own cluttered office.

Whether in astronomy, genome sequencing, or high energy physics, Revson's Hubble experience with scientific communication is replicated in offices and laboratories around the country. In this chapter, we address a number of issues in scientific communication that have been profoundly changed by the new information technology. In earlier chapters, we first looked at the Garvey/Griffith model, the paper-based system mapped during the 1960s, that established steps of the scientific communications system. We then identified important changes during the interim 30 years and developed four models based upon technological and socioeconomic projections. These include electronic communication systems with no journals and with or without peer review. Finally, we observed information handling in three scientific specialties today; now we will examine changes, using the baseline model of Garvey and Griffith.

The Garvey/Griffith Model Today

The original Garvey/Griffith model was developed during a "print-on-paper" era. The model described scientific communication broadly across many disciplines. It captured the scientific communication system that had evolved gradually over the previous century, and continued to reflect communication behavior for some thirty years afterwards. The refereed scientific article was the key element in the system; and both formal and informal communication led to journal publication as the expected outcome of scientific research.

The system served additional functions including support for a reward structure that had long been the norm for scientists working in higher education. It also fostered the development of the present array of scientific publishers that includes both not-for-profit associations and commercial organizations who produce both primary journals and the secondary indexing and abstracting services that facilitate access to the contents of journals.

More recently, organizational and societal changes have created a dynamic environment that exerts pressure for new ways to examine issues and new approaches to scientific problem-solving. Big Science has become "Bigger Science" with increased emphasis on collaborative and team research directed to address interdisciplinary problems of global importance, and whose solutions require expensive facilities and equipment and create enormous data sets. Concurrently, computer-based information technologies have emerged that are beginning to change the ways scientists use, produce, and disseminate information. In this work we have examined three scientific specializations—the space sciences, genome research, and high energy physics—that have appropriated technology to devise new systems and structures to support communication.

The specializations herein studied have experienced particularly rapid changes and have adapted technologies in innovative ways. The motivations that impelled scientists in these fields to shift to new modes of communication are likely complex. In our case study approach, we have identified a number of variables that seem to be influential, and we hypothesize that further changes will occur in other fields where some of the same characteristics prevail. From our examinations of the three fields described in earlier chapters, we can identify these change agents:

- high level of research front activity, which fosters a need for very rapid communication of research
- reliance on informal communication which leads to a high level of invisible college interaction
- large-scale collaborative projects involving dispersed research teams
- multi-institutional research projects at scattered locations
- large data sets that can be shared and used for multiple experiments.

Each of these change agents can stimulate shifts in communication behaviors that, in turn, lead to new models such as those presented here. We examine the traditional Garvey/Griffith model and identify where changes might be expected.

At the beginning of a research project, informal communication is important to scientists who are developing methodologies and refining hypotheses. Colleagues at invisible colleges may be queried for details on construction of experimental apparatus or for data on related objects of study. In an earlier time, telephone calls and visits to other laboratories as well as conference interactions provided opportunities for communication. The ready availability of e-mail now makes such direct communication even easier, and likely faster and less costly, than a telephone call or a visit.

An additional potential source of information is provided by open listserves—a question posted there goes beyond an individual's known colleagues and taps a wider pool of expertise. (We should note, however, that not all lists are open; some are closed with all subscription requests vetted by the listserve owners, which serves to limit participation.) Such a group might be considered an "electronic invisible college," defined by the participants and unrestricted by virtue of whom one knows. Participants only need to have suitable computer and communication equipment, and access to appropriate lists or news groups.

As a project nears completion, its researchers prepare manuscripts describing the findings. As indicated in the section on manuscript preparation, now virtually all manuscripts are produced on word processors and thus exist in electronic form from their inception. Still other research projects, such as the Human Genome Project, generate data that are transferred to central depositories. These may be viewed as types of "publication" that make information available earlier than if a refereed journal article were the means of distribution. Posting of gene sequencing data in a shared repository diverges from the past practice of including data in a refereed journal article; it is noteworthy here that data represent the unit of distribution.

All of these innovations—listserves, preprint archives, and shared depositories of data—have changed communication in significant ways:

- invisible colleges may be more accessible and more freely joined
- research results are available sooner and in unrefereed forms
- articles and data are the units of distribution rather than journal issues
- authors are becoming direct publishers through World Wide Web pages
- boundaries are blurring between informal and formal communication.

The changes cited above are beginning to impact journal publishers. They are responding with their own innovations, primarily in modes of distribution. The well-established system of peer review may also adapt to an electronic environment in response to these changes. We discuss these and other aspects of scientific publishing in the next section.

Publishing: the Shift From Print

A research undertaking in most scientific fields generally follows a predictable pattern. A scientist or group of scientists has an idea, develops a methodology to test the idea, collects and analyzes relevant data, and draws conclusions from the analysis. Having completed the experiment, the scientist prepares a manuscript, explaining results of the study in context with the literature on similar studies; submits the manuscript to a peer-reviewed journal; revises it as recommended by reviewers and editors; and, assuming the manuscript is accepted, has the satisfaction of seeing the article in print. Prior to publication, usually an editor, two or three reviewers, and some colleagues had critically evaluated the study and its findings.

With the publication of this permanent record, an important component of the scientific process begins. It is only after publication and the resulting widespread distribution that the scientific community has the opportunity to determine a particular study's importance and its permanence in the realm of all other scientific discoveries and advances.

What is the potential of electronic publishing to alter this time-honored scenario of communicating scientific information? An examination of how electronic publishing has already changed and influenced the present publications system foretells an even greater impact on the future of scientific communication.

Manuscript Preparation and Submission

Those individuals with access to a computer have shown an almost universal, rapid acceptance of word processing for manuscript preparation. Prior to its advent, revisions of one sentence, let alone whole documents, were burdensome and time consuming. In less automated times, authors simply ignored many corrections and alterations to text because of the difficulty of making corrections.

Moreover, the electronic transfer of information has great advantages. A manuscript can be transmitted via a local area network (LAN), electronic mail, or the Internet to co-authors, colleagues, or members of an electronic invisible college; a statistical program can be run at an off-site computer; results can be downloaded in table format directly into text; graphics, charts, diagrams; laboratory findings can be transmitted from a distant computer directly to the document; illustrations can be scanned and reproduced; and citations can be compiled in database management software and selectively added to an article. These types of transfers are relatively common; they not only greatly assist in manuscript preparation, but they lessen the chance of transcribing errors. Today, most journal editors require (or at least expect) that along with a paper copy of a submitted manuscript, the author submit a disk containing an exact electronic version of the text, diagrams, and tables.

Editorial Peer Review Process

Before they are disseminated, scientific results are usually subjected to the rigors of the editorial peer review process. Even if not universally successful, editorial peer review offers a solid attempt to validate scientific results before publication. The review filters out the redundant, the incorrect, and the useless information, and gives the "imprimatur of scientific authenticity" (Ziman 1968, 111) to a research project and its conclusions. While the electronic environment may significantly alter the editorial peer review process, it is essential to maintain some form of validation.

We should point out that electronic manuscripts do present logistical problems. The document will need to be printed, which raises issues of compatible equipment, downloading or ftp-ing text, correct formatting (especially for charts, diagrams, or illustrations), and transcription of special characters (diacritics, mathematical symbols, and super- and subscripts, for example). Material in color is particularly problematic. Perhaps the most serious issue is confidentiality. Anonymity is simply not compatible with an electronic environment.

Even if the editorial review process remains virtually unchanged, its advantages in an electronic environment are substantial. These advantages include the following:

- eliminates the time and expense of photocopying and mailing

- gives reviewers the opportunity to carry on a dialog with the editor, author(s), and other reviewers
- makes raw data available to aid in the evaluation
- allows the editor and reviewers to enter comments directly into the text
- shortens the time between research results and publication.

Whenever the process of editorial peer review is altered such that documents routinely appear on an electronic network without prior evaluation or review, the validation of the material in those documents then becomes the responsibility of the whole scientific community after general distribution. In a commentary in the *British Medical Journal,* LaPorte et al (1995) suggested this very scenario. They proposed "a truly democratic system" (p. 1388) of reviewing: readers of electronic articles serve as reviewers; each reader would have the opportunity to add to a "comment card;" each article would be given a "priority score;" and the relative worth of an article would be determined by the number of times an article was retrieved (an electronic "impact factor").

The cited commentary presented a thought-provoking piece to generate comments and to suggest a rethinking of scientific communication and the validation of research. The American Chemical Society, expressing concerns about altering editorial peer review in an electronic environment, distribution of scientific information, and the quality of the information, stated that "the mandate remains clear: the archives of science must contain all the quality research within the discipline that is validated for the record by peer-review and publication" (American Chemical Society 1995).

Are there examples of journals that provide a prototype for peer review in an electronic environment? Since 1978 Stevan Harnad has edited *Behavioral and Brain Sciences* (BBS), a journal that provides "open peer review commentary to researchers." (Harnad 1995a). In BBS, the traditional editorial peer review process is used to decide which manuscripts to publish. Only after the manuscript is accepted is it "circulated to a large number of commentators selected by the editors, referees, and author to provide substantive criticism, interpretation, elaboration, and pertinent commentary and supplementary material" A selection process determines who is eligible to make commentaries: "qualified professionals . . . if they have (1) been nominated by current BBS associates, (2) refereed for BBS, or (3) had a commentary or article accepted for publication." (ibid) While BBS has provided a type of letters to-the-editor by publishing commentaries at the end of the article, it has not provided an alternate to the traditional peer review system.

Several online electronic journals have addressed editorial peer review issues. The first peer-reviewed electronic medical journal, the *Online Journal of Current Clinical Trials*, began in 1992. The editor, Edward Huth (1992), confirmed that all reports and reviews would be peer reviewed. The announcement of the publication of the *Online Journal of Knowledge Synthesis for Nursing* included a statement that "manuscripts will be peer reviewed so there is assurance of the same quality one finds in paper journals." (Barnsteiner 1993) Stevan Harnad expressed the same views in describing editorial peer review in the online journal *Psycholoquy*, which he edits (Harnad 1995b). Kassirer (1992), editor of the *New England Journal of Medicine*, echoed this view: "we may have to redefine what we mean by a manuscript [but] we must preserve the independent scrutiny of our methods, results, and interpretations."

Although altering the process of editorial peer review in an electronic environment has been explored, little has changed in practice. Also, strong sentiment exists that it should not change. The proliferation and corresponding problems of listserves provide insight into some of the issues that might be encountered with an open electronic peer review system. How would one judge the knowledge or correctness of a contributor; how do readers sort the useful from the useless information; how (or should) one limit input of those readers who comment on almost every issue; when will a reader know if a solution has been reached and a conclusion drawn?

Finally, if a document is subjected to numerous revisions, how would one identify the corrected or most current version versus the original document? No one person would be responsible for its final validation. In the physics preprint file, for example, linking the preprint to the published document is the responsibility of authors as well as the staff of the Stanford Linear Accelerator (SLAC) Library Database. Approximately one-half of the preprints have a reference to a published document (Galic 1996). In all likelihood some links to published high energy physics articles will never be made.

Many examples of identifying post-publication errors exist, both intentional and unintentional (LaFollette 1992). It is clear that, without a system of built-in checks and balances, errors would be more difficult to detect and correct than they now are. The National Library of Medicine has developed and implemented a system to correct errors and to identify retractions (Kotzin and Schuyler 1989), which could serve as a model for an electronic environment.

The role of peer review within the electronic environment will continue to be discussed and evaluated. The degree to which it might be altered or changed depends upon the value the scientific community places on the evaluative process. Whatever the final outcome, a permanent, archival record of a validated work remains most important.

Publication

Technology offers the means of bypassing the printed publication process. If the publisher is circumvented in an electronic environment, it raises questions about the collection, sale, and distribution of scientific information. The publisher plays a critical role in all of these steps and has added an important value to the publication process. Publishers help to organize scientific knowledge by subject. In every field, certain publishers are known to be more selective, more scholarly, or more able to attract the most progressive thinkers. Screening mechanisms and value-added components provided by publishers could be incorporated into the production of electronic documents. A perusal through the Internet reveals numerous attempts to organize information that underscore the difficultly and mixed results of such undertakings.

Electronic Journals

Electronic journals take on a range of formats: online version only, CD-ROM format, and a mix of formats including print. This variation in formats will continue for sometime and may even become more mixed. No complete data exists on the number or percentage of scientific journals that have made the transformation to electronics or on the level of transformation that has taken place.

The *British Medical Journal* (BMJ) has announced that it is available on the Internet (Delamothe 1995). An accompanying editorial stated that BMJ's home page was restricted to "text because images, sound, and video take a long time to download with currently available software" (p. 1343). As of January, 1996 the Internet version of BMJ contained the table of contents, abstracts of articles, and full text of some editorials. With few exceptions this offers little advantage over a standard indexing or abstracting service. The decision to convert *Encyclopaedia Britannica* to a continuously updated, online text (Reid 1995) has transformed the encyclopedia, considered by some to be obsolete, to a remarkably flexible, easy-to-use reference resource. The preprint database, described in Chapter 4, could have a substantial impact on the type or format of high energy physics journals. The most dramatic result would be a decision to cease publishing the printed version of these texts in physics journals. The preprint file would then become the archival record.

We find many examples of projects to make the full text of already published journals available online or on CD-ROM:

- American Chemical Society—*Chemical Journals Online* (ACS Publications on Command 1995)
- Stanford Libraries/American Society of Biochemistry and Molecular Biology—*Journal of Biological Chemistry Online*, including text, tables and graphs (Stanford Libraries 1995)
- American Psychiatric Electronic Library (APEL) (Epstein 1995)
- Red Sage project (Butter 1994)
- Ovid project, which began as a tool to access MEDLINE and expanded to include the full-text of medical journals (Rogers 1995)

The number of fully electronic journals has grown considerably The fifth edition of the *Directory of Electronic Journals, Newsletters, and Academic Discussion Lists* by the Association of Research Libraries (1995) includes 675 journals and newsletters. The first edition, published in 1991, listed 110 journals and newsletters. According to a special report in *Science*, as of the end of 1995, more than one hundred peer reviewed science, technical, and medical journals exist on the Internet (Taubes 1996).

As long as electronic journals are thought of as merely mirror images of a paper copy, they will have limited use. Electronic publications have the potential to link numerous types of information and to add a wealth of enhancements, among them, full text of the cited material, raw data, detailed explanations or descriptions, electronic discussions following the distribution of a manuscript, and authors' home pages. Large electronic files of raw data linked to electronic journals might necessitate the continuance of print-based publications to explain, draw conclusions, and evaluate the data.

Electronic journals also have the capability for updating work continuously by adding data and altering content as new information becomes available. An important value of electronic journals is the ability to move beyond text, tables, and graphics and to link the reader to interactive programs, videos, audio displays, and radiological images. Examples include tracking weather patterns, performing computations on data, viewing a rotating three dimensional figure, monitoring a chemical synthesis, examining details of a surgical brain procedure, monitoring the effect of a drug on blood flow through the brain, or determining the aerodynamic implication of a new wing design for an airplane.

Nonetheless, some questions remain to be answered concerning interactive electronic journals:

- If a manuscript or document is dynamic, what is the "archival" version?
- Which is the copyrighted version?
- Who is responsible for updating text?
- Who is (are) the author(s)?

Standards

Standards are essential for transmitting accurate data, text, and graphics. Hickey (1995) described a number of the standards available today for electronic journals and the advantages and disadvantages of each. As long as text and simple tables predominated in electronic material, the ASCII (American Standard Code for Information Interchange) standard was sufficient. A recreation of the original format, with a corresponding ability to search and manipulate, requires a more sophisticated standard.

SGML (Standard Generalized Markup Language) is an international standard of importance to electronic publishing. HTML (Hypertext Markup Language) is a specific form of SGML suitable for an Internet environment to control the display of a document. These standards have many variations and a final solution has yet to emerge. For example, one of the goals of the Human Genome Project is to link massive amounts of data in different formats. A major challenge is to integrate these databases. One solution being tried is a "middleware" program that would connect a federation of databases (Five years of progress 1995).

The traditional model of organizing journals by volume, issue, and page numbers may not be sufficient or appropriate in an electronic environment. Methods of identifying and labeling electronic publications need to be developed. The need to group a journal issue arbitrarily by more-or-less unrelated articles no longer pertains. An arrangement of articles by subject within each journal issue is an option.

For the most part, electronic journal formats remain similar to the paper version. Any corrections needed after a manuscript has been published appear in the next electronic version of the journal. The errors continue in both the electronic and paper version of the original article. The decision to correct an error in a subsequent publication is using a "paper-based solution" in an electronic setting.

Economics of Publishing

Many economic issues remain to be played out. Who will ultimately pay for electronic access remains to be resolved. When journals are available in several formats, each format adds a cost to journal production. King and Griffiths (1995) identified four types of electronic publications and reviewed the economic implications of each: full-text journals online, image files on CD-ROM, truly electronic journals, and scholarly writing published by universities. They predicted that universities will play an increasingly important role in electronic publishing of the faculty work. University libraries are an obvious vehicle to take on the responsibility of managing access to electronic publications and, as an extension, to play an increasingly more important role in the publication process. For-profit publishers will probably become more competitive and professional societies may take on more responsibility, with a fee structure for use.

In all likelihood, a move toward more restricted access to Internet sites will take place as the need to make a profit drives policy issues. As of this writing, the full-text online version of the *Journal of Biological Chemistry* offers unlimited access through Stanford University's pilot project; if the project continues, some restrictions to access will likely be imposed. Software, hardware, set-up fees, and licensing fees that cost thousands of dollars give access to the material, not ownership of it.

Initially, the *Online Journal of Current Clinical Trials* could be accessed only through a standalone CD-ROM player with dedicated software at a cost of about three dollars per unit. It was produced by the most logical means at a time prior to wide Internet availability and illustrates how quickly the 'Net has changed access to information. New solutions are now being tried, and it is currently one of a group of journals managed by OCLC's FirstSearch using Guidon software. The BMJ's home page gives the journal more visibility, but does not generate income for the journal or provide enough information to replace a subscription.

In making a decision to put material on the Internet, one must evaluate numerous other considerations, not the least of which is the revenue generated by the sale of publications. The provider can, of course, impose a password restriction with corresponding charging that provides a mechanism for accessing any point on the Internet. Revenue generated by access or by printing documents would probably be insufficient to finance the cost of producing a journal. Much of the value-added elements of published print is the information in advertisements. And, of course, advertising provides an important revenue-producing source, especially for commercial organizations. Advertising income is a key element in the publications equation that must be resolved for electronic publishing.

Publishing and Communication

A news item in *Nature* announced that a sequence of 900,000 nucleotide bases is available on the Internet to help researchers identify a breast cancer gene (Dickson 1995). This is an excellent example of the benefits of global electronic communication, not only one-way, but two-way communication.

To draw an analogy with Kuhn (1962), periods of confusion accompany paradigm shifts. When an anomaly first occurs, those affected attempt to make the anomaly "lawlike." Electronic publishing or, more generally, electronic communication, can be viewed as an anomaly. Initially, users attempted to communicate electronically in the same way as on paper. Editors required both paper and disk copies of manuscripts, and e-mail messages are printed out to read or save. Online library catalogs were initially formatted exactly like the printed catalog cards, and electronic journals still have volume numbers and issue numbers. We conjecture, however, that eventually the whole process of scientific communication will be transformed into one that is unrecognizable from the current practice.

Information Needs and Use

When text, images, and sound are digitized and moved across electronic networks, how scientists seek information and communicate with each other are radically transformed. Proximity and location are no longer issues when information can be transmitted to desktop computers and when the user can call up a laboratory and print out 43 pages of data without taking a trip from Chicago to Los Alamos. In the example of the Hubble Telescope and other observatories, scientists share instrumentation at other sites and direct instructions to the operators on how and what they wish to observe.

In traditional hard copy publishing, finding information entailed identifying a physical document that contained it, usually in a library. Electronic digitization enables the development of databanks, situated where the information originates or in digital libraries that have electronic connections among concepts and objects. This creates an "information space" that the user can explore to find interlinked information (Schatz 1991-92). What is sought may be an object (e.g., genes or chromosomes), an abstraction without a physical analog (e.g., limits and continuity), a process (e.g., algorithms). Ultimately, a user will be able to trace a chain of ideas or information through a series of files without regard for its location (Pool 1993). Interface agents or "knowbots" will assist users in coordinating different tasks by determining the parameters of a search, formulating queries, and selecting relevant information. The information can be downloaded or delivered to the user from the source.

Much information exchange is made through "invisible colleges," in which scientists working on problems in related areas informally talk about their work, consult with each other, and test new ideas (Price 1986). The C. Elegans Electronic Community System (Schatz 1991-92) and The Upper Atmosphere Research Collaboratory in Greenland (University of Michigan 1995), both initiatives of the National Science Foundation, are test beds for the new electronic invisible college. In the digital world, the invisible colleges have taken electronic form so that participants in a project, spread across many locations, can converse much as if they were sitting at the same conference table.

In such an environment, users need to know how to navigate the digital world or have the support to do so. The resources are continually evolving and fast-moving: users will need training and updating to use new formats, to access the Internet, and to develop search strategies for locating network-based information. Clearly, those scientists who have the capability or the support for accessing the range of media, for sharing instrumentation in other locations, for locating software to process data, and for working in teams, will enjoy the advantage over their colleagues.

In the prototype C. Elegans Electronic Community System, researchers will be able to share information and insights by adding their observations, comments, and findings to the record. Beyond controlling instruments remotely and accessing information from numerous databases, Schatz sees the project as offering the potential for scientists to collaborate and use information in new ways (Pool 1993).

Information Professionals and Organizations: the Future of the Library

In commenting on the speed of advances in science today, Koshland (1990) cites Sidney Brenner "that a modern computer hovers between obsolescence and the nonexistent." Brenner goes on to say that one of the most unsettling features of science is the rapidity at which it changes our cultural values and institutions.

What are the enduring values and functions of organizations that handle information? Scientists can now do their own searches, order the data they need through an electronic network, and use software to create and distribute publications. As digital libraries become widespread, the physical location of materials will become less relevant and libraries will no longer exist in their present form.

General publishers (e.g., McGraw-Hill, Elsevier, University Microfilms) and professional societies (e.g., American Astronomical Society) are already producing their scientific journals in dual format—traditional printed copy and electronic versions. They will increasingly distribute their products (journals, articles, graphics, and data) online so that they can eliminate the expense and bother of warehousing those items and delivering them to bookstores and libraries. Users will be able to search and obtain full length articles directly from the originator or from the publisher. *The Astrophysical Journal Letters* now exists completely online, linked to a database, so that astronomers can click on references and see abstracts or full texts—30 days before the printed version is available (Navigating the Net 1995).

Navigating a constantly changing electronic information system is a complex undertaking. Its practice has become a specialty that requires hands-on experience and current awareness of new developments. New roles have developed for information professionals and libraries to fill the needs of both support services and education.

Weingarten (1996) suggests five roles for information professionals and libraries in the National Information Infrastructure: (1) provide access to leading edge information services, as these are not always available or affordable and are continuously changing; (2) provide basic support services, as there are many barriers—technical, economic, and intellectual—in setting up facilities to use the Web; (3) serve as navigator/guide to sophisticated tools as the National Information Infrastructure evolves; (4) serve as archivist/depository/authenticator because electronic information is extremely perishable and severe problems occur in preserving original and interim documentation; and (5) organize information spaces among the commercial, public, private, and government sectors to assure access to information.

The evolution to an electronic system is perhaps most accelerated in fast moving fields of science that have research fronts and the infrastructure already in place. The Association of Research Libraries cautions that, for widespread use, to move vast amounts of data at a high speed requires an adequate three-tier structure: a national backbone, regional networks, and campus or local area networks. The library is now viewed as one element of the present communications system—the publisher, the printed book, the monograph, the process of peer review, and the copyright system will be affected (Okerson 1996). On the present transitional state, Peck (1996) comments that "at no time in the history of scholarly publishing have the social structures and organizational patterns of creating and maintaining the scholarly word had such serious examination."

The Shift from Print: How Soon?

This book has presented a case for the ultimate transformation of scientific communication. We suggest that it will, and should, evolve from a print-based system dominated by the refereed scientific journal to an electronic system in which the basic units of distribution of information may well be individual articles and data. We believe that the numerous advantages offered by a networked-based structure facilitate both informal and formal communication among scientists, and these will eventually transform the present system.

The evolution will be gradual; some early changes being merely electronic versions of paper-based communication. The slow pace of change can be seen in some of the electronic journals that endeavor to look exactly like their printed page counterparts, even employing the same type face and page layout. Ultimately, we believe that new formats will emerge that, unlike their paper predecessors, could be dynamic, interactive, multimedia, nonlinear, and more. Network-based "publication" may be initiated by authors producing documents that might look very much like those found in preprint databases. And peer review will likewise adapt to an electronic environment, taking on a more open form by involving more scientists whose commentaries can be read and evaluated by others. An e-print of the future will possess features of both informal and formal formats now in use.

We believe that print and electronic information sources are likely to co-exist for some time to come, although major changes in the scientific publishing sector will probably occur. Libraries and librarians will have new opportunities to participate in the evolving communications system that will transform roles while at the same time preserving some of their traditional functions.

Karen Drabenstott (1994) has described a "library of the future" that is increasingly a digital library:

> For a while, library users will rely on the paper collections that libraries have amassed over the years. As information is increasingly produced in digital artifact form . . . paper collections will

slowly fall into disuse and large portions of such collections will be warehoused at remote locations. . . . Libraries will continue to be associated with buildings. Although physical collections of books, journals and other materials will no longer consume valuable space in these buildings, we can envision the need for workspaces where users consult state-of-the-art computer workstations; study spaces where users demand quiet for contemplation and reflection. Although computer and communications technology provides the foundation for the library of the future, in its current state of development technology also erects some barriers. Future challenges include intersystem compatibility and standardization, maintaining data integrity and security, high band-width requirements for non-text information transfer, and more.

Nonetheless, technological issues are likely to be resolved more easily than the economic, legal, organizational, and sociological problems that the inevitable changes will bring. Their solutions will require the cooperation of many diverse types of organizations in finding areas in which common interests and goals permit collaboration. As one observer puts it:

> There is, without doubt, a conflict of interest among the stakeholders in the current system of scholarly communication. Today we have a balance of these conflicting interests that is working less effectively each day. Finding a new balance will require both cooperation and (constructive) confrontation. [Lynch 1992, 108]

A recent commentary in *Science* (Noam 1995) discusses how electronic production and distribution of information undermines the traditional flow of information and with it the structure of major institutions of learning. With the development of alternative teaching tools and the capability for long-distance learning from outstanding scholars, Noam foresees many functions of universities being superseded and their role in intellectual inquiry greatly reduced.

Larger societal implications are underscored by Rush (1996), who projects effects on egalitarianism/elitism, employment, nature of the workplace, and quality of life. The focus, according to Rush, is about change and the likely consequences of change for every individual. Similar debates have arisen from other technologies over the centuries: consider the printing press, the telephone, and nuclear power. But the most daunting aspect, he continues, is the inability of the individual, organization, or even nation to exercise a significant degree of control over it.

From print to electronic—the views and implications are well summed up by Merton (1942). Scientists are concerned with the norms and values of science, legitimized by their own group, but they are also dependent upon the social structure in which they live and work. Information technology will not easily change such values as objectivity, organized skepticism, accountability to one's peers, or the reward system that provides recognition and esteem. The continued search to ensure quality control in the electronic database of the Human Genome Project and in the signals collected by space instruments reflects this ethic. Investigators will continue detached scrutiny of theory and beliefs, pending examination of data. Nor will society with its "corresponding obligations and interests" in ownership, competition, and status be basically altered by the new technology. However, these interests may undergo changes in magnitude and be better served for fewer. Some consequences we have already seen are accelerated competition, ecological changes in institutions and work forces, and greater maldistribution of resources. On the other hand, for scientists, it is clear that information technology provides invaluable support for investigation by bringing changes in speed, accuracy, convenience, and ability to connect ideas. This is what we see happening in the disciplines we have observed.

That we should further pursue the fallout in personal and societal outcomes from information technology has been suggested, such as roles of the various players—scientists, network organizations, computer specialists, publishers, information professionals, and other users (Small, 1996). Griffith (1996) notes that "the broad system of science awards an incredibly fragile commodity, namely creativity and original discovery." In this study, we have shown how technology has transformed the way scientists work, but we do yet know how this affects their scientific creativity and output. The Mertonian perspective—the effect on the scientific ethos, on institutionalized interests, and on the practice of science itself—have yet to be investigated in assessing this evolution from a paper-based to a digital system.

References

American Chemical Society. 1995. *Will science publishing perish? The paradox of contemporary science journals.* ACS Publications. American Chemical Society.

ACS Publications on Command. 1995. An electronic library of all American Chemical Society journals at your fingertips. http://pub.acs.org/acselec/oncom.html.

Association of Research Libraries. 1995. *Directory of electronic journals, newsletters and academic discussion lists.* 5th edition. Washington, DC: Association of Research Libraries, Office of Scientific and Academic Publishing.

Barnsteiner, J.H. 1993.. The Online Journal of Knowledge Synthesis for Nursing. *Reflections,* Spring, 19(1):8.

Butter, K. A. 1994. Red Sage: the next step in delivery of electronic journals. *Medical Reference Services Quarterly*, Fall, 13:75-81.

Delamothe, T. 1995. BMJ on the internet. Visit our home page at http://www.bmj.com/bmj/. *British Medical Journal*, May 27, 31(6991):1343-4.

Dickson, D. 1995. Open access to sequence data "will boost hunt for breast cancer gene." *Nature*, Nov. 3, 378:425.

Drabenstott, K.M. 1994. *Analytical Review of the Library of the Future*. Washington DC: Council on Library Resources.

Epstein, B. A. 1995. American Psychiatric Electronic Library: an initial evaluation. *Medical Reference Services Quarterly*, Spring, 14(1):1-8.

Five years of progress in the human genome project. 1995. *Human Genome News*, September-December, 4-9.

Flam, F. 1993. Space telescope institute: inside the Black Box. *Science*, June 18, 260:1716-17.

Galic, H. 1996. Personal communication.

Griffith, B.C., communication with author. Drexel University; Philadelphia PA, May 1996.

Harnad, S. 1995a. Editorial policy. *Behavioral and Brain Sciences*, Dec., 18(4):i.

———— 1995b. Implementing peer review on the net: Scientific quality control in scholarly electronic journals. In Peck, R. and G. Newby, eds. *Electronic publishing confronts academia: The agenda for the year 2*. Cambridge MA: MIT Press.

Hickey, T. B. 1995. Present and future capabilities of the online journal. *Library Trends*, Spring, 43:528-43.

Huth, E.J. 1992. The Online Journal of Current Clinical Trials. *New England Journal of Medicine*, April 3, 326(18):1227.

Kassirer, J.P. 1992. Journals in bits and bytes. Electronic medical journals. *New England Journal of Medicine*, January 16, 326(3):195-7.

Kotzin, S. and P.L. Schuyler. 1989. NLM's practices for handling errata and retractions. *Bulletin of the Medical Library Association*, October, 77(4):337-42.

King, D. W. and J. Griffiths. 1995. Economic issues concerning electronic publishing and distribution of scholarly articles. *Library Trends*, Spring, 43(4):713-4.

Koshland, D. 1990. To see ourselves as others see us. *Science*, Jan. 5, 247:9.

Kuhn, T.S. 1970. The Structure of scientific revolutions. Chicago, University of Chicago Press.

LaFollette, M.C. 1992. *Stealing into print: fraud, plagiarism, and misconduct in scientific publishing*. Berkeley, CA: University of California Press.

LaPorte, R. E., E. Marler, S. Akazawa, F. Sauer, C. Gamboa, C. Shenton, C. Glosser, A. Villasenor, and M. Maclure. 1995. The death of the biomedical journals. *British Medical Journal*, May 27, 31(6991):1387-9.

Lynch, C. 1992. Reaction, response, and realization: from the crisis in scholarly communication to the age of networked information. *Serials Review*, Spring/Summer, 18, No. 1/2.

Merton, R. 1942. The normative structure of science. In *The sociology of science: theoretical and empirical investigations*. Chicago, Univ. of Chicago Press, 267-278.

Navigating the Net. 1995. *Science*, December 22, 270:1903.

Noam, E. M. 1995. Electronics and the dim future of the university. *Science*, October 13, 270(523):247-9.

Okerson, A. 1996. University libraries and scholarly communication. In Peek, R.B. and G. Newby, eds. *Scholarly publishing: the electronic frontier*. Cambridge, MA, MIT Press: 3-15.

Peek, R.B. 1996. Scholarly publishing: facing the new frontiers. In Peek, R.B. and G. Newby, eds. *Scholarly publishing: the electronic frontier*. Cambridge, MA, MIT Press: 3-15.

Pool, R. Beyond databases and e-mail. 1993. *Science*. August 13, 261: 843.

Price, D. 1986. *Little science, big science . . . and beyond*. New York, Columbia University Press, 56-81.

Reid, C. 1995. Britannica Online debuts on the Web. *Publishers Weekly*, October 16, 242:15.

Rogers, M. 1995. Ovid technologies networks for NYC medical facilities; Cornell, Memorial Sloan-Kettering, Rockefeller, and the Hospital for Special Surgery to share T3 line. *Library Journal*, November 1, 12:21.

Rush, J.E. 1996. Forward. In Peek, R.B. and G. Newby, eds. *Scholarly publishing: the electronic frontier*. Cambridge, MA, MIT Press: 3-15.

Schatz, B.R. 1991-92. Building an electronic community system. *J. of Management Information Systems*, Winter, 8, 87-107.

Small H., communication with author. *Institute for Scientific Information*, Philadelphia, PA, May 1996.

Stanford Libraries. 1995. *Journal of Biological Chemistry*. http://WWW-jbc. stanford.edu/jbc/.

Taubes, G. 1996. Science journals go wired. *Science*, February 9, 271:764-766.

University of Michigan. Upper Atmospheric Research Collaboratory. 1995. *Design Considerations in building a distributed collaboratory*. Ann Arbor, MI: University of Michigan School of Library and Information Studies.

Weingarten, F.W. 1996. Five great roles for Libraries and Librarians within the NII. *American Libraries*, January, 27:17.

Ziman, J.M. 1968. *Public Knowledge: An essay concerning the social development of science*. London: Cambridge University Press.

Index